Recent Titles in This Series

491 **Tom Farrell and Lowell Jones,** Markov cell structures near a hyperbolic set, 1993
490 **Melvin Hochster and Craig Huneke,** Phantom homology, 1993
489 **Jean-Pierre Gabardo,** Extension of positive-definite distributions and maximum entropy, 1993
488 **Chris Jantzen,** Degenerate principal series for symplectic groups, 1993
487 **Sagun Chanillo and Benjamin Muckenhoupt,** Weak type estimates for Cesaro sums of Jacobi polynomial series, 1993
486 **Brian D. Boe and David H. Collingwood,** Enright-Shelton theory and Vogan's problem for generalized principal series, 1993
485 **Paul Feit,** Axiomization of passage from "local" structure to "global" object, 1993
484 **Takehiko Yamanouchi,** Duality for actions and coactions of measured groupoids on von Neumann algebras, 1993
483 **Patrick Fitzpatrick and Jacobo Pejsachowicz,** Orientation and the Leray-Schauder theory for fully nonlinear elliptic boundary value problems, 1993
482 **Robert Gordon,** G-categories, 1993
481 **Jorge Ize, Ivar Massabo, and Alfonso Vignoli,** Degree theory for equivariant maps, the general S^1-action, 1992
480 **L. Š. Grinblat,** On sets not belonging to algebras of subsets, 1992
479 **Percy Deift, Luen-Chau Li, and Carlos Tomei,** Loop groups, discrete versions of some classical integrable systems, and rank 2 extensions, 1992
478 **Henry C. Wente,** Constant mean curvature immersions of Enneper type, 1992
477 **George E. Andrews, Bruce C. Berndt, Lisa Jacobsen, and Robert L. Lamphere,** The continued fractions found in the unorganized portions of Ramanujan's notebooks, 1992
476 **Thomas C. Hales,** The subregular germ of orbital integrals, 1992
475 **Kazuaki Taira,** On the existence of Feller semigroups with boundary conditions, 1992
474 **Francisco González-Acuña and Wilbur C. Whitten,** Imbeddings of three-manifold groups, 1992
473 **Ian Anderson and Gerard Thompson,** The inverse problem of the calculus of variations for ordinary differential equations, 1992
472 **Stephen W. Semmes,** A generalization of riemann mappings and geometric structures on a space of domains in $\mathbf{C}^n$, 1992
471 **Michael L. Mihalik and Steven T. Tschantz,** Semistability of amalgamated products and HNN-extensions, 1992
470 **Daniel K. Nakano,** Projective modules over Lie algebras of Cartan type, 1992
469 **Dennis A. Hejhal,** Eigenvalues of the Laplacian for Hecke triangle groups, 1992
468 **Roger Kraft,** Intersections of thick Cantor sets, 1992
467 **Randolph James Schilling,** Neumann systems for the algebraic AKNS problem, 1992
466 **Shari A. Prevost,** Vertex algebras and integral bases for the enveloping algebras of affine Lie algebras, 1992
465 **Steven Zelditch,** Selberg trace formulae and equidistribution theorems for closed geodesics and Laplace eigenfunctions: finite area surfaces, 1992
464 **John Fay,** Kernel functions, analytic torsion, and moduli spaces, 1992
463 **Bruce Reznick,** Sums of even powers of real linear forms, 1992
462 **Toshiyuki Kobayashi,** Singular unitary representations and discrete series for indefinite Stiefel manifolds $U(p,q;\mathbb{F})/U(p-m,q;\mathbb{F})$, 1992
461 **Andrew Kustin and Bernd Ulrich,** A family of complexes associated to an almost alternating map, with application to residual intersections, 1992
460 **Victor Reiner,** Quotients of coxeter complexes and P-partitions, 1992

(Continued in the back of this publication)

Number 491

Markov Cell Structures near a Hyperbolic Set

Tom Farrell
Lowell Jones

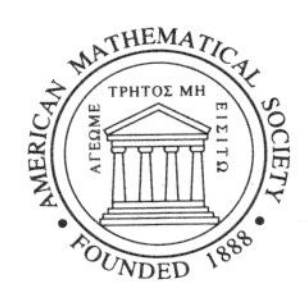

May 1993 • Volume 103 • Number 491 (second of 4 numbers) • ISSN 0065-9266

American Mathematical Society
Providence, Rhode Island

1991 *Mathematics Subject Classification.*
Primary 58F15, 58F12, 58F09; Secondary 57R05, 57R30, 57R50.

Library of Congress Cataloging-in-Publication Data

Farrell, Tom, 1929–
Markov cell structures near a hyperbolic set/Tom Farrell, Lowell Jones.
p. cm. – (Memoirs of the American Mathematical Society, ISSN 0065-9266; no. 491)
"Volume 103, number 491 (second of 4 numbers)."
Includes bibliographical references.
ISBN 0-8218-2553-4
1. Diffeomorphisms. 2. Manifolds. 3. Hyperbolic spaces. I. Jones, Lowell, 1945– . II. Title. III. Series.
QA3.A57 no. 491
[QA613.65]
510 s–dc20 93-464
[514′.72] CIP

Memoirs of the American Mathematical Society

This journal is devoted entirely to research in pure and applied mathematics.

Subscription information. The 1993 subscription begins with Number 482 and consists of six mailings, each containing one or more numbers. Subscription prices for 1993 are $336 list, $269 institutional member. A late charge of 10% of the subscription price will be imposed on orders received from nonmembers after January 1 of the subscription year. Subscribers outside the United States and India must pay a postage surcharge of $25; subscribers in India must pay a postage surcharge of $43. Expedited delivery to destinations in North America $30; elsewhere $92. Each number may be ordered separately; *please specify number* when ordering an individual number. For prices and titles of recently released numbers, see the New Publications sections of the *Notices of the American Mathematical Society.*

Back number information. For back issues see the *AMS Catalog of Publications.*

Subscriptions and orders should be addressed to the American Mathematical Society, P. O. Box 1571, Annex Station, Providence, RI 02901-1571. *All orders must be accompanied by payment.* Other correspondence should be addressed to Box 6248, Providence, RI 02940-6248.

Memoirs of the American Mathematical Society is published bimonthly (each volume consisting usually of more than one number) by the American Mathematical Society at 201 Charles Street, Providence, RI 02904-2213. Second-class postage paid at Providence, Rhode Island. Postmaster: Send address changes to Memoirs, American Mathematical Society, P. O. Box 6248, Providence, RI 02940-6248.

Printed in the United States of America.
This volume was printed directly from author-prepared copy.
The paper used in this book is acid-free and falls within the guidelines established to ensure permanence and durability. ♾
10 9 8 7 6 5 4 3 2 1 98 97 96 95 94 93

CONTENTS

Abstract

Let $F : M \longrightarrow M$ denote a self-diffeomorphism of the smooth manifold M and let $\Lambda \subset M$ denote a hyperbolic set for F. Roughly speaking a Markov cell structure for $F : M \longrightarrow M$ near Λ is a finite cell structure C for a neighborhood of Λ in M such that for each cell $e \in C$ the image under F of the unstable factor of e is equal to the union of the unstable factors of a subset of C, and the image of the stable factor of e under F^{-1} is equal the union of the stable factors of a subset of C.

The main result proven in this paper is that for some positive integer q the diffeomorphism $F^q : M \longrightarrow M$ has a Markov cell structure near Λ.

A precise statement of this result and an outline of its proof can be found in the introduction of this paper.

A list of open problems related to Markov cell structures and hyperbolic sets can be found in section 15. Section 15 can be read independently of sections 2–14.

1. INTRODUCTION*

In this section the main result of the paper is formulated and an outline of its proof is given. The main result states that any diffeomorphism $F : M \to M$ has a Markov cell structure near any of its hyperbolic sets $\Lambda \subset M$. Roughly speaking a Markov cell structure for $F : M \to M$ near Λ consists of a cell structure for a neighborhood of Λ, and a positive integer q, such that the behavior of each cell under the application of $F^q : M \to M$ imitates the behavior of a partition set in a Markov partition for $F^q : \Lambda \to \Lambda$.

This result was announced by the authors in [6].

The authors have proved the existence of Markov cell structures for expanding maps in dimension two in [7]. By referring to [7] the reader can see what motivated the much more complicated arguments of this paper.

STATEMENT OF RESULTS

Let $F : M \to M$ denote a C^1 diffeomorphism of the smooth manifold M onto itself. By a **hyperbolic set** for $F : M \to M$ we mean a compact subset $\Lambda \subset M - \partial M$ which satisfies the following properties.

<u>1.1</u> (a) $F(\Lambda) = \Lambda$.

(b) There is a neighborhood U of Λ in M such that $\Lambda = \bigcap_{i=-\infty}^{\infty} F^i(U)$.

(c) Λ is equipped with a hyperbolic structure. That is there are subbundles ξ_u, ξ_s of $T(M)$ and there are numbers $a \in (0,1)$, $\lambda > 1$ all satisfying the following. Both of ξ_u, ξ_s are left invariant by $df : T(M) \to T(M)$; $T(M)|_\Lambda = \xi_u \oplus \xi_s$; for each $v \in \xi_u$ and each integer $q > 0$ we have $|dF^q(v)| > a\lambda^q|v|$; for each $v \in \xi_s$ and each integer $q > 0$ we have $|dF^q(v)| < a^{-1}\lambda^{-q}|v|$. (Here, for any vector $v \in T(M)$, we let $|v|$ denote the norm of v with respect to a given Riemannian metric on M.)

A cell complex C for a region $|C|$ of M is called a **regular cell complex** (or a **regular cell structure**) if the attaching map $f_e : \partial B^k \to C^{k-1}$ for each cell $e \in C$ is an embedding onto a subcomplex of the $(k-1)$-skeleton C^{k-1}, where $k = \dim(e)$ and B^k is the unit ball in R^k. We shall use the term **k-cell** (or **cell**) to denote any topological space which is homeomorphic to the unit k-ball B^k.

---

*Each author was supported in part by the NSF.
Received by editor June 24, 1988.

A regular cell structure C for a region of M is termed a **rectangular cell structure near** Λ if the following properties hold.

1.2. (a) $|C|$ is a neighborhood for Λ in M.

(b) Set n, m equal to the dimensions of the fibers in the bundles ξ_u, ξ_s. For each $e \in C$ there must exist an embedding $h_e : R^n \times R^m \to M$ and cells $e_u \subset R^n$, $e_s \subset R^m$ such that $h_e(e_u \times e_s) = e$.

(c) For each $e \in C$ set $\partial_u e = h_e(\partial e_u \times e_s)$ and set $\partial_s e = h_e(e_u \times \partial e_s)$. Then each of $\partial_u e$ and $\partial_s e$ must be a subcomplex of C.

For C as in 1.2 and any non-negative integers i, j we denote by C^i_j the union $\cup(e - \partial e)$ where this union runs over all $e \in C$ such that $\dim(e_u) \leq i$ and $\dim(e_s) \geq j$.

DEFINITION 1.3. *A* **Markov cell structure** *for $F : M \to M$ near the hyperbolic set $\Lambda \subset M$ consists of a finite rectangular cell structure C near Λ, and a positive integer q, satisfying the following properties.*

(a) $F^q(C^i_j) \cap |C| \subset C^i_j$ holds for all non-negative integers i, j.

(b) Let $e_0, e_1, e_2, \ldots, e_i$ denote a collection of cells in C. Denote by A^+ the closure of the set of all points $x \in M$ such that $F^{qj}(x) \in e_j - \partial e_j$ holds for all $j \in \{0, 1, 2, \ldots, i\}$, and denote by A^- the closure of the set of all points $x \in M$ such that $F^{-qj}(x) \in e_j - \partial e_j$ holds for all $j \in \{0, 1, 2, \ldots, i\}$. Then there must be subsets $A^-_s \subset R^m$ and $A^+_u \subset R^n$ which satisfy the following properties.

(i) If $A^- \neq \emptyset$ then A^-_s is a topological cell with $\dim(A^-_s) = \dim((e_i)_s)$. If $A^+ \neq \emptyset$ then A^+_u is a topological cell with $\dim(A^+_u) = \dim((e_i)_u)$.

(ii) $h_{e_0}(A^+_u \times (e_0)_s) = A^+$ and $h_{e_0}((e_0)_u \times A^-_s) = A^-$. (See 1.2 for h_{e_0}.)

REMARK 1.4. Property 1.3(a) is the most important property of a Markov cell structure. Note that 1.3(a) implies that $F^{-q}(\hat{C}^j_i) \cap |C| \subset \hat{C}^j_i$, where $\hat{C}^j_i$ denotes the union $\cup(e - \partial e)$ with this union running over all cells $e \in C$ such that $\dim(e_u) \geq i$ and $\dim(e_s) \leq j$. Recall that each set in a Markov partition for $F : \Lambda \to \Lambda$ must be a product $P_u \times P_s$ in terms of the canonical charts imposed on Λ by the stable foliation and unstable foliation, and $F : \Lambda \to \Lambda$ preserves this local product structure parametrized by the canonical charts (see [5], [25]). Unfortunately neither the stable foliation not the unstable foliation for Λ need fill up an entire neighborhood for Λ in M; so there is no natural local product structure for M in a neighborhood of Λ. We hypothesize 1.3(b) to make up for this deficiency.

We can now state the main theorem of this paper.

THEOREM 1.5. *There is a neighborhood U for Λ in M and there is a positive number N. For any $\varepsilon > 0$ there is a Markov cell structure C for $F : M \to M$ near Λ satisfying the following properties.*

(a) $U \subset |C|$.

(b) Each cell $e \in C$ has diameter less than ε.

(c) We may choose q in 1.3 to be any positive integer which satisfies $q \geq N$.

REMARK 1.6. It might be said that we prove our theorem 1.5 from scratch in that we do not refer in our proofs to any predecessor theorems. However we are following in spirit the proofs of many predecessor theorems. We remark now on the relationship of our theorem to some of these predecessor theorems. Adler and Weiss [1] (linear maps in dimension two) and Sinai [25] (in all dimensions) have proven the existence of Markov partitions for Anosov diffeomorphisms. If $F : M \to M$ of 1.5 is an Anosov diffeomorphism (i.e., $\Lambda = M$) then the top dimensional cells of a Markov cell structure for $F : M \to M$ are the partition sets of a Markov partition for $F : M \to M$. Bowen has extended the work of Adler and Weiss [1], and Sinai [25], by proving that each basic set of an Axiom A diffeomorphism has a Markov partition (cf. [5]). If $F : M \to M$ of 1.5 is a Axiom A diffeomorphism and Λ denotes one of its basic sets (note that every basic set is a hyperbolic set in the sense of 1.1) then it should be possible to choose a Markov cell structure C for $F : M \to M$ near Λ such that the intersection of Λ with the maximal cells of C yields a Markov partition for $F : \Lambda \to \Lambda$. The preceding claim is not yet proven, however there is the following technique to get a Markov partition for $F : \Lambda \to \Lambda$ from a Markov cell structure C for $F : M \to M$ near Λ. Let P be the collection of all subsets of Λ gotten as follows: for each cell $e \in C$ let e_Λ denote the interior of $(e - \partial e) \cap \Lambda$ within the space Λ, let e'_Λ denote the closure of e_Λ within the space Λ, and set $P = \{e'_\Lambda : e \in C\}$. Note that P satisfies all the properties of being a Markov partition for $F^q : \Lambda \to \Lambda$ with the following possible exception: two distinct sets $S_1, S_2 \in P$ are allowed to intersect on their interiors. For such a collection of sets there is a simple recipe which generates from P a Markov partition for $F^q : \Lambda \to \Lambda$ (cf. [5], [25]). A basic tool used by all of these authors to construct Markov partitions is the canonical charts imposed on Λ by the stable and unstable foliations. The canonical charts used by Adler and Weiss come from linear algebra, those used by Sinai come from the work of Anosov [2], and those canonical charts used by Bowen come from the work of Hirsch, Kupka, Pugh, Smale (cf. [11], [25]). Generally speaking the canonical charts do not fill up a neighborhood of the hyperbolic set Λ. So we can not base our constructions on canonical charts, but rather must use what are approximately canonical charts (see section 14 of this paper).

OUTLINE FOR THE PROOF OF THEOREM 1.5.

We give here an outline of chapters 2 through 14, which together make up the proof of theorem 1.5.

So that the reader does not get lost in technical details we approach this outline by first sketching a proof of the existence of Markov cell structures for an expanding immersion $F : M \to M$ on a smooth closed compact manifold M. After this we describe where the proof of theorem 1.5 departs from this sketch and the complications involved in those departures.

Recall that $F : M \to M$ is called an **expanding immersion** if it is a C^1 immersion which satisfies $|df^i(v)| > a\lambda^i|v|$ for all $v \in T(M)$, for some numbers $a \in (0,1)$ and $\lambda > 1$ which are independent of v and i.

By a **Markov cell structure** for the expanding immersion $F : M \to M$ we mean a regular cell structure C for M together with a positive integer q such that the following is satisfied.

1.7. $F^q(C^k) \subset C^k$ for all $k \in \{0, 1, \ldots, m = \dim(M)\}$.

The reader should compare 1.7 to 1.3(a). Note that 1.7 is essentially the property that 1.3(a) imposes on the "unstable factor."

THEOREM 1.8. *Any expanding immersion $F : M \to M$ has a Markov cell structure.*

SKETCH FOR THE PROOF OF THEOREM 1.8.

A **smooth polyhedron** P in M consists of a partition of a region $|P|$ in M into a finite number of subsets of M which satisfy the following properties.

1.9 (a) Each $e \in P$ is a connected compact piecewise smooth submanifold of M with boundary ∂e.

(b) For each $e \in P$ $(e - \partial e)$ is a smooth submanifold of M.

(c) For any two distinct $e, e' \in P$ we have that $(e - \partial e) \cap (e' - \partial e') = \emptyset$. Moreover $e \cap e'$, ∂e, $\partial e'$ are all equal to finite unions of members of P.

If each member $e \in P$ is a smooth simplex in M then we call P a **smooth triangulation** for the region $|P|$ of M. Denote by $D(P)$ the minimum distance between any two disjoint members of P. For each member $e \in P$ denote by $d(e)$ the diameter of e, and set $\tau(P) = \min\{D(P)/d(e) : e \in P\}$. Roughly speaking, $\tau(P)$ is an approximate lower bound for the angles which occur in the partitioning P.

The theory of piecewise smooth triangulations tells us there is a number $b > 0$, and for each $\varepsilon > 0$ there is a smooth triangulation K for all of M, satisfying the following (cf. [18]).

1.10 (a) $\tau(K) > b$, $D(K) < \varepsilon$.
(b) Each simplex in K is "almost" linear.

A **first derived subdivision** of K, denoted $K^{(1)}$, is a smooth triangulation of M satisfying the following properties.

1.11 (a) $K^{(1)}$ is a subdivision of K.
(b) The vertices of $K^{(1)}$ consist of those of K together with one additional vertex in the interior of every simplex of K.

An **rth derived subdivision** of K, denoted by $K^{(r)}$, is obtained by taking a first derived subdivision of some $K^{(r-1)}$.

Note that it follows from 1.10 that for any positive integer r there is a number $b_r > 0$, and for any $\varepsilon > 0$ there are smooth "almost" linear triangulations K, L for M, which satisfy the following properties.

1.12 (a) K is an rth derived subdivision of L.
(b) L satisfies $\tau(L) > b$, $D(L) < \varepsilon$.
(c) $\tau(K) > b_r$; $D(K)/D(L) > b_r$.

By a **ball structure** for K we mean a collection of piecewise smooth m-balls $\{Y(e) : e \in K\}$ in M ($m = \dim(M)$) which satisfy the following.

1.13 (a) For any subcomplex X of K we have that $\cup_{e \in X} Y(e)$ is a neighborhood for $|X|$ in M.
(b) For any $e, e' \in K$ we have that $Y(e) \cap e' \neq \emptyset$ if and only if $e \subset e'$; and $Y(e) \cap Y(e') \neq \emptyset$ if and only if either $e \subset e'$ or $e' \subset e$.
(c) Let $e_1, e_2, \ldots, e_n$ be simplices in K such that $e_i \subset e_{i+1}$ holds for all $i = 1, 2, \ldots, n-1$. Choose subsets $J, J' \subset \{1, 2, \ldots, n\}$ such that $J \cap J' = \emptyset$ and $J \cup J' = \{1, 2, \ldots, n\}$. Choose $e \in K$ with $e_n \subset e$. Then the intersection $e \cap (\cap_{j \in J} Y(e_j)) \cap (\cap_{j \in J'} \partial Y(e_j))$ must be a piecewise smooth cell of dimension equal to $\dim(e) - |J'|$, where $|J'|$ equals the cardinality of J'. Moreover if $J \neq \emptyset$ then the intersection $\partial e \cap (\cap_{j \in J} Y(e_j)) \cap (\cap_{j \in J'} \partial Y(e_j))$ is a piecewise smooth cell of dimension equal $\dim(e) - |J'| - 1$.

By a **redundant ball structure** for K (of order k) we mean k distinct ball structures $\{Y_i(e) : e \in K\}$, $i = 1, 2, \ldots, k$, for K which satisfy the following properties.

1.14 (a) For each $e \in K$ randomly choose one of the $\{Y_i(e) : i = 1, 2, \ldots, k\}$ and denote this choice by $Y(e)$. Then the $\{Y(e) : e \in K\}$ must be a ball structure for K.

(b) For any $e \in K$, and any $1 \le i < j \le k$, we have either $Y_i(e) \subset Y_j(e)^o$ or $Y_j(e) \subset Y_i(e)^o$—where X^o = interior(X).

The remaining part of the sketch of the proof for theorem 1.8 consists of the five claims made in the next five steps.

STEP I. There is a number $\bar{b}_r > 0$ (independent of ε in 1.12) such that for any smooth triangulation K for M as in 1.12 there is a redundant ball structure $\{Y_i(e) : e \in K,\ i = 1, 2, \dots, m\}$ of order m for K which satisfies the following properties (where $m = \dim M$).

1.15 (a) Each $Y_i(e)$ is an almost linear polyhedron in M. Locally each $Y_i(e)$ is (almost) linearly equivalent to the transversal intersection of half spaces.

(b) All of the smooth polyhedra $\{Y_i(e) : e \in K,\ i = 1, 2, \dots, m\}$ and K are in general position (i.e. transverse position) to one another in M.

(c) Let P denote the partition of M generated by the smooth polyhedra $\{Y_i(e) : e \in K,\ i = 1, 2, \dots, m\}$ and K. Then P must be an almost linear polyhedron in M.

(d) $\tau(P) > \bar{b}_r$; $D(P)/D(K) > \bar{b}_r$.

We owe the reader two definitions in 1.15(b)(c). A collection of smooth polyhedra $\{K_i : i \in I\}$ in M are in **transverse position** to one another in M if for every selection of members $e_i \in K_i,\ i \in I$, we have that the smooth submanifolds $\{e_i - \partial e_i : i \in I\}$ of M are in transverse position to one another in M.

If $\{K_i : i \in I\}$ is a finite collection of smooth polyhedra in M then the **partition P of M generated by the** $\{K_i : i \in I\}$ is defined as follows: a subset of $e \subset M$ is a member of P if and only if e is equal to the closure of a maximal non-empty intersection of sets in the collection $\{e_i - \partial e_i, M - e_i : e_i \in K_i,\ i \in I\}$.

STEP II. (Triangulating image balls).

There is a number $N > 0$ which is independent of ε in 1.12. There is a positive integer r which depends only on $m = \dim(M)$. Suppose that the integers r of 1.12 and 1.15 are the same as the integer r given in this step. Then for each integer $q > N$ there is a homeomorphism $h : M \to M$ which satisfies the following properties, for sufficiently small ε in 1.12(b).

1.16. (a) For each $x \in M$ the distance from x to $h(x)$ is less than $a^{-1}\lambda^{-q}\varepsilon$.

(b) For each $e \in K, i \in \{1, 2, \dots, m\}$, the mapping $F^q \circ h : Y_i(e) \to M$ is an embedding of $Y_i(e)$ onto a subcomplex of K.

For any ball $Y_i(e)$, we set $Y_i'(e) = h(Y_i(e))$.

STEP III. (Thickening balls).

Note that it follows from 1.13 and 1.14 that all the boundaries $\{\partial Y_i(e) : e \in K,\ i = 1, 2, \ldots, m\}$ are in transverse position to one another in the piecewise smooth category. In particular we have the following satisfied.

1.17 Any $m+1$ distinct boundaries in $\{\partial Y_i(e) : e \in K,\ i = 1, 2, \ldots, m\}$ have an empty intersection.

For each $e \in K, i \in \{1, 2, \ldots, m\}$, let $K(e, i)$ denote the subcomplex of K having $F^q \circ h(\partial Y_i(e))$ for its underlying set (see 1.16(b)). For each simplex $d \in K(e, i)$ choose one of the balls in $\{Y_j(d) : j = 1, 2, \ldots, m\}$, and denote by $H(e, i)$ the collection of all such chosen balls. Note that it follows from 1.17 that the collections $H(e, i)$ can be made so as to satisfy the following property.

1.18 For any $e, e' \in K$, and any $i', i \in \{1, 2, \ldots, m\}$, if we have that $\partial Y_i(e) \cap \partial Y_{i'}(e') \neq \emptyset$ then we must have that $H(e, i)$ and $H(e', i')$ have no balls in common.

For each positive integer t and for each of the balls $Y_i(e)$ we can now define the **tth level thickening** if $Y_i(e)$—denoted by $Y_i(e; t)$—as follows. For each $Y_j(d) \in H(e, i)$ let $Y_j''(d)$ denote the component of the pre-image $F^{-q}(Y_j'(d))$ which intersects $Y_i'(e)$. Set $Y_i(e; 1)$ equal to the union of $Y_i'(e)$ with all the balls $Y_j''(d)$ such that $Y_j(d) \in H(e, i)$. To get $Y_i(e; t)$ we assume that all the $\{Y_j(d; t-1) : d \in K,\ j = 1, 2, \ldots, m\}$ have already been defined. Let $H^{t-1}(e, i)$ denote the collection of all $Y_j(d; t-1)$ such that $Y_j(d) \in H(e, i)$, and for each such $Y_j(d, t-1)$ let $Y_j'(d, t-1)$ denote the component of the pre-image $F^{-q}(Y_j(d; t-1))$ which intersects $Y_i'(e)$. Set $Y_i(e; t)$ equal to the union of $Y_i'(e)$ with all the balls $Y_j'(d; t-1)$ such that $Y_j(d; t-1) \in H^{t-1}(e, i)$.

We can now state the main claim of this step.

1.19 For each positive integer t there is a homeomorphism $h_t : M \to M$ such that $h_t(Y_i(e)) = Y_i(e; t)$ holds for all $e \in K$ and all $i \in \{1, 2, \ldots, m\}$.

The reader might be wondering by now why we deal with redundant ball structures, rather than just sticking to ball structures. The answer is that 1.19 could not possibly be true for ball structures. To see this we suppose that $\dim(M) = 2$ and that we are given only the one ball structure $\{Y_1(e) : e \in K\}$ for K (instead of a redundant ball structure of order two). Consider two balls $Y_1(e')$ and $Y_1(e)$ which intersect as in figure 1.20(a). Since we have only the one ball structure the sets $H(e, 1)$ and $H(e', 1)$ can not be chosen to satisfy 1.18. In fact $H(e, 1)$ and

$H(e', 1)$ must have exactly the two balls $Y_1(d_1)$ and $Y_1(d_2)$ in common, where d_1, d_2 are the two points of $F^q \circ h(\partial Y_1(e)) \cap F^q \circ h(\partial Y_1(e'))$. Thus the sets $Y_1(e;1)$ and $Y_1(e';1)$ must intersect as in figure 1.20(b), where the heavy solid lines of 1.20(b) indicate the intersection of the boundaries $\partial Y_1(e;1) \cap \partial Y_1(e';1)$.

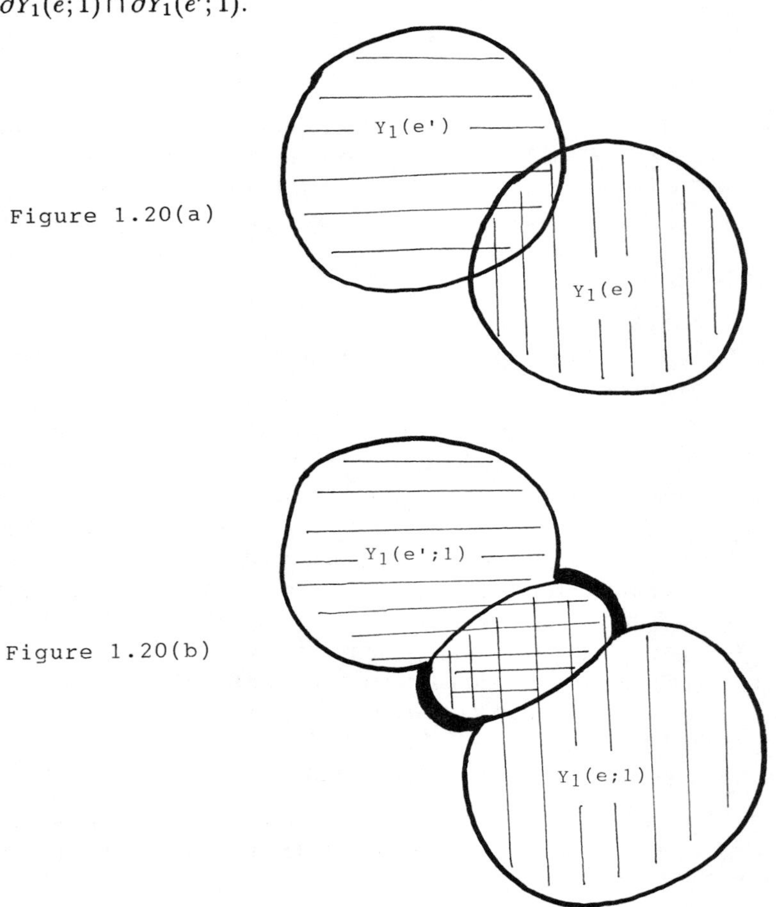

Figure 1.20(a)

Figure 1.20(b)

STEP IV. (The limit balls).

For each ball $Y_i(e)$ we define $Y_i(e;\infty)$ to be the closure in M of the union $\cup_{j \geq 1} Y_i(e;j)$. We call $Y_i(e;\infty)$ the limit ball for $Y_i(e)$. The claim of this step is the following.

<u>1.21</u> There is a homeomorphism $h_\infty : M \to M$ such that $h_\infty(Y_i(e)) = Y_i(e;\infty)$ holds for all $e \in K$ and all $i \in \{1, 2, \ldots, m\}$.

An obvious start towards proving 1.21 would be to set h_∞ equal to $\lim_{t\to\infty} h_t$, where the $h_t : M \to M$ are the homeomorphisms of 1.19. We can assure convergence of this limit by insisting that the h_t satisfy in addition to 1.19 the following property.

<u>1.22</u> $h_t \circ h_{t-1}^{-1}(F^{-q(t-1)}(Y_i(e)) = F^{-q(t-1)}(Y_i(e;1))$ holds for all $t \geq 1$ and for all $e \in K$, $i \in \{1,2,\ldots,m\}$.

A limit argument, based on the equality $h_t(Y_i(e)) = Y_i(e;t)$ of 1.19 and on 1.22, will now show that $h_\infty(Y_i(e)) = Y_i(e;\infty)$ must hold for all $e \in K$, $i \in \{1,2,\ldots,m\}$. So to complete the verification of 1.21 it remains only to show that $h_\infty : M \to M$ is a homeomorphism. This may not be true unless some further tinkering with the $h_t : M \to M$ is carried out. The details of the "tinkering" can be found in section 12.

STEP V. (Construction of Markov cell structures).

There is another partition of M, denoted by Q, to which the $\{Y_i(e) : e \in K, i = 1,2,\ldots,m\}$ give rise. A subset $A \subset M$ will be a member of the partition Q if and only if A is equal to the closure of a maximal non-empty intersection of sets in the collection $\{Y_i(e), \partial Y_i(e), M - Y_i(e) : e \in K, i = 1,2,\ldots,m\}$. Note that it follows from 1.13 and 1.14 (see in particular 1.13(c)) that

<u>1.23.</u> Q is a regular cell structure for M.

Define C to be the image of Q under the homeomorphism $h_\infty : M \to M$ of 1.21. It is now just a straightforeward exercise to deduce from 1.16, 1.21 and from the definition of C, that F and C satisfy 1.7.

This completes the sketch for the proof of theorem 1.8.

We now discuss how the proof of theorem 1.5 differs from the proof of theorem 1.8 just given, and where the details of such differences are found in the main body of this paper. These differences are discussed under each of the following seven titles.

FOLIATION HYPOTHESIS. Roughly speaking the proof of 1.5 consists of applying the arguments in the proof of 1.8 first to the unstable directions of $F^q : M \to M$ near the hyperbolic set Λ, and then applying these same arguments to the unstable directions of $F^{-q} : M \to M$ near Λ. We make the following simplifying assumption in sections 2 through 13 (see 5.2): there are smooth foliations $\mathcal{F}_u, \mathcal{F}_s$ of a neighborhood V of Λ in M such that at Λ $\mathcal{F}_u$ is tangent to ξ_u and $\mathcal{F}_s$ is tangent to ξ_s; for any leaves L_u, L_s in $\mathcal{F}_u, \mathcal{F}_s$ we have that $F(L_u) \cap V$, $F(L_s) \cap V$ are open subsets of leaves of $\mathcal{F}_u, \mathcal{F}_s$. In section 14 we show how to remove this hypothesis. Note that this foliation hypothesis implies that there are smooth canonical coordinates for $F : M \to M$ near Λ which fill up an entire neighborhood for Λ in M. These canonical coordinates give to a neighborhood of Λ a smooth local product structure: one factor of this local product structure is the unstable direction of $F : M \to M$; and the other factor is the stable direction of $F : M \to M$. In section 14 we show how to replace the smooth canonical coordinates by "approximate" canonical coordinates which give rise to an "approximate" local product structure near Λ.

TRIANGULATIONS. From this point on let us concentrate on the constructions which need to be made in the unstable direction of $F : M \to M$ near Λ. Unlike the proof of 1.8, where K was chosen to be a smooth triangulation for all of M, we now choose K to be a finite collection of smooth triangulations "in the unstable direction" of $F : M \to M$ near Λ which fit together nicely. In more detail let $g_i : R^n \times R^m \to M, i \in I$, denote a finite number of smooth canonical charts which fill up a neighborhood of Λ. Then roughly speaking K is a collection of smooth triangulations $\{K_i : i \in I\}$ of arbitrarily large compact regions $|K_i|$ of R^n (where R^n corresponds to the unstable direction of $F : M \to M$ near Λ) such that within the intersection Image$(g_i) \cap$ Image(g_j) either the partitioning $g_i(K_i \times R^m)$ subdivides the partitioning $g_j(K_j \times R^m)$ or vice-versa for all $i, j \in I$.

The precise statement of the properties which the collection K should satisfy is given in Proposition 6.8. The proof of the existence of such a collection K requires definitions and constructions given in sections 2,3,4.

REDUNDANT BALL STRUCTURES.

In the proof of 1.8 we considered a redundant ball structure $\{Y_i(e) : e \in K, i = 1, 2, 3, \ldots, m\}$ for the smooth triangulation K, where $m = \dim(M)$. Now in the proof of 1.5 we must consider for each of the smooth triangulations K_i, $i \in I$, a redundant ball structure $\{Y_j(e) : e \in K_i, j = 1, 2, \ldots, x\}$ for K_i. We require that $x = \beta(n+m+1)n(10\eta)^{2m}$ where η, β come from 6.1, 8.2, $n = \dim(\xi_u)$, $m = \dim(\xi_s)$ (see 8.2 and 9.2). We also require that all the redundant ball structures $\{Y_j(e) : e \in K_i, j = 1, 2, \ldots, x\}$, $i \in I$, fit together nicely. A precise statement of the conditions which all these ball structures are required to satisfy is given in lemma 7.2. The proof of the existence of such a collection of redundant ball structures relies on definitions and constructions given in sections 2,3,4.

TRIANGULATING IMAGES OF BALLS.

We need an analogue for the proof of 1.5 of step II in the proof of 1.8. For technical reasons which we won't go into at this time it is necessary before stating this analogue to refine the redundant ball structures

$\{Y_j(e) : e \in K_i, j = 1, 2, \ldots, x\}, i \in I$. This is done by choosing for each $i \in I$ a covering of a compact region of R^m by balls $\{B_{s,j} : j \in I_i\}$ where I_i is a finite index set depending on $i \in I$. A precise statement of the conditions which the coverings $\{B_{s,j} : j \in I_i\}, i \in I$, are required to satisfy is given in 8.1. We then choose for each $i \in I$ and each $j \in I_i$ a subcollection of the balls $\{Y_k(e) : e \in K_i,\ k = 1, 2, \ldots, x\}$—which is denoted by $\{_jY_p(e) : e \in K_i, p = 1, 2, \ldots, y\}$. We require that this subcollection also be a redundant ball structure for K_i of order y where $y = (n + m + 1)n(10\eta)^{2m}$ (see 9.2). A precise statement of all the conditions which the redundant ball structures $\{_jY_p(e) : e \in K_i,\ p = 1, 2, \ldots, y\},\ i \in I$, are required to satisfy is given in lemma 8.2.

Now we can state the analogue for the proof of 1.5 of 1.16 in the proof of 1.8. Roughly speaking we require that for any $i, i' \in I,\ j \in I_i,\ e \in K_i$, and $p \in \{1, 2, \ldots, y\}$ if we have that $F^q(g_i(_jY_p(e) \times B_{s,j})) \subset g_{i'}(|K_{i'}| \times (\cup_{z \in I_{i'}} B_{s,z}))$ then the image of $F^q(g_i(_jY'_p(e) \times B_{s,j}))$ under the composition map

$$M \xrightarrow{g_{i'}^{-1}} R^n \times R^m \xrightarrow{\text{proj.}} R^n$$

is the underlying set of a subcomplex of the smooth triangulation $K_{i'}$. Here $_jY'_p(e)$ is the image of $_jY_p(e)$ under a homeomorphism which is C^0 close to the identity map (and which depends only on i, j). A precise statement of this requirement is given in 8.5.

THICKENING BALLS.

We need an analogue for the proof of 1.5 of step III in the proof of 1.8. That is for each $i \in I,\ j \in I_i,\ e \in K_i,\ p \in \{1, 2, \ldots, y\}$, and for each positive integer t we must define the tth level thickening of $_jY_p(e)$—denoted by $_jY_p(e; t)$. A precise description of how these thickenings are defined is given in section 9. The thickening theorem 9.6 is the analogue of 1.19.

Here is a rough description of this thickening process. For any positive number r let $_rB_{s,j}$ denote the ball of radius rr_j which has the same center point in R^m as does $B_{s,j}$, where r_j denotes the radius of the ball $B_{s,j}$. For each $i, i' \in I,\ j \in I_i,\ e \in K_i, j' \in I_{i'},\ p \in \{1, 2, \ldots, y\}$, we define a subcomplex $K(j, p, e; j')$ of $K_{i'}$ as follows. If $F^q(g_i(_jY_p(e) \times {}_2B_{s,j}))$ is contained in $g_{i'}(|K_{i'}| \times ({}_{3/2}B_{s,j'} \cap (\cup_{z \in I_{i'}} B_{s,z})))$, then $K(j, p, e; j')$ is the subcomplex of $K_{i'}$ whose underlying set is the image of $F^q(g_i(\partial_j Y_p(e) \times {}_2B_{s,j}))$ under the composition map

$$M \xrightarrow{g_{i'}^{-1}} R^n \times R^m \xrightarrow{\text{proj.}} R^n.$$

Otherwise $K(j,p,e;j')$ is the empty subcomplex. For each simplex d in $K(j,p,e;j')$ we choose one of the balls $\{_{j'}Y_{p'}(d) : p' = 1,2,\ldots,y\}$ and denote by $H(j,p,e;j')$ the collection of all such chosen balls. The collections $H(j,p,e;j')$ are the analogues to the collections $H(e,i)$ in step III of the proof of 1.8. There is also an analogue (for the $H(j,p,e;j')$) of the condition 1.18 imposed on the $h(e,i)$ (see lemma 9.2). Now for each $_jY_p(e)$ and each $H(j,p,e;j')$ set $_{j',j}Y_p(e;1)$ equal the subset of R^n which is mapped under the composite

$$R^n \times 0 \xrightarrow{g_i} M \xrightarrow{F^q} M \xrightarrow{g_{i'}^{-1}} R^n \times R^m \xrightarrow{\text{proj.}} R^n$$

onto the union of all the sets $_{j'}Y'_{p'}(d)$ such that $_{j'}Y_{p'}(d)$ is in $H(j,p,e;j')$. (See "Triangulating Images of Balls" for $_{j'}Y'_{p'}(d)$.) Define $_jY_p(e;1)$ to be the union of $_jY'_p(e)$ with all the $_{j',j}Y_p(e;1)$ (here j,p,e are fixed and j' is varying). The higher level thickenings $_jY_p(e;t)$ are defined by induction. Suppose that all the $\{_kY_b(d;t-1)\}$ have already been defined. Let $H^{t-1}(j,p,e;j')$ denote the collection of all $_kY_b(d;t-1)$ such that $_kY_b(d)$ is a member of $H(j,p,e;j')$. Then if we replace each $H(j,p,e;j')$ by $H^{t-1}(j,p,e;j')$ in the description of $_jY_p(e;1)$ given above we will get $_jY_p(e;t)$ instead of $_jY_p(e;1)$.

LIMIT BALLS.

For each ball $_jY_p(e)$ we define the limit ball $_jY_p(e;\infty)$ to be the closure of the union $\cup_{t\geq 1}\, _jY_p(e;t)$. The limit theorem 12.1 is the analogue for the proof of 1.5 of 1.21.

MARKOV CELL STRUCTURES.

We need an analogue for the proof of 1.5 of step V in the proof of 1.8. This is carried out in section 13. Here is a rough description of the constructions of section 13. We must first repeat all the constructions listed under the previous six titles for the map $F^{-1} : M \to M$ instead of for the map $F : M \to M$. We then get triangulations $\{L_i : i \in I\}$ for compact regions of R^m (instead of for R^n), and limit balls $\{_jY_p(e;\infty) : e \in L_i, j \in J_i, p = 1,2,\ldots,y\}$ for each $i \in I$ which lie in R^m (instead of in R^n). To get the Markov cell structure C we choose a special subset S of the collection $\{g_i(_jY_p(e;\infty) \times {_{j'}Y_{p'}}(e';\infty)) : i \in I, j \in I_i, j' \in J_i, e \in K_i, e' \in L_i, p,p' \in \{1,2,\ldots,y\}\}$ and define C to be the cell structure "generated" by the sets in S under the operations of taking boundaries, complements and intersections.

This completes our outline of the proof of theorem 1.5.

§2. SOME LINEAR CONSTRUCTIONS

This section contains the statement of two propositions (2.10, 2.14) which deal with piecewise linear triangulation theory. The proofs of these propositions are given in section 3. Piecewise smooth versions of these propositions are stated in section 4.

This section also contains a great deal of notation and terminology which is used through the rest of the paper.

DEFINITION 2.1. *A* **linear polyhedron** *K in R^n consists of a partition of a region $|K|$ in R^n into a finite number of subsets $\{e \in K\}$ satisfying:*

(a) Each $e \in K$ in a connected compact P.L. submanifold of R^n with boundary ∂e.

(b) Each $e \in K$ is contained in a k-dimensional plane of R^n where $k = \dim(e)$.

(c) For any two distinct $e, e' \in K$ we have that $(e - \partial e) \cap (e' - \partial e') = \emptyset$. Moreover $e \cap e', \partial e, \partial e'$, are all equal finite unions of members of K.

A subset $L \subset K$ is called a **subcomplex** of the linear polyhedron K if L is a linear polyhedron. We denote by K^i the subcomplex of K consisting of all members of K which have dimension less than or equal to the non-negative integer i.

For any $e \in K$ we let e denote both a subset of R^n and the subcomplex of K which has the subset $e \subset R^n$ as the underlying set. We denote by $|K|$ the union $\bigcup_{e \in K} e$.

A linear polyhedron K in R^n is called **full** if $\overline{|K| - |K^{n-1}|} = K$.

A linear polyhedron K is a **linear triangulation** if each $e \in K$ is a linear simplex.

Recall that a **subdivision** of the linear polyhedron K consists of another linear polyhedron L in R^n such that $|K| = |L|$ and each $e \in K$ is a subcomplex of L.

DEFINITION 2.2. *A subdivision L of a linear triangulation K is called a* **derived subdivision** *if L is a linear triangulation and for each $e \in K$ there is exactly one vertex of L in $e - \partial e$.*

DEFINITION 2.3. *Let K denote a linear triangulation in R^n having for vertices the collection $\{v_i : i \in I\}$. Let $\{g_i : [0,1] \to R^n : i \in I\}$ denote paths in R^n with $g_i(0) = v_i$ for all $i \in I$. Suppose for each $t \in [0,1]$ the points $\{g_i(t) : i \in I\}$ are the vertices of another linear triangulation K_t such that the map $\varphi_t : K^0 \to K_t^0$ given by $\varphi_t(v_i) = g_i(t)$ for all $i \in I$ extends to a simplicial isomorphism $\varphi_t : K \to K_t$. Then the maps $\varphi_t : K \to K_t$, $t \in [0,1]$, are called a* **P. L. flow** *of K.*

DEFINITION 2.4. *We call the subdivision L of the linear triangulation K an* r-**fold derived subdivision** *of K if L is obtained by first applying a derived subdivision process to K r times in succession to get a subdivision T of K, and then choosing a P. L. flow $\varphi_t : T \to T_t$, $t \in [0,1]$, which satisfies $\varphi_t(e) = e$ for all $t \in [0,1]$ and all $e \in K$, and setting $L = T_1$. Any r-fold derived subdivision of K is denoted by $K^{(r)}$.*

DEFINITION 2.5. *Let $\{K_i : 1 \leq i \leq u\}$ denote a finite collection of full linear polyhedra in R^n. The partition W which is* **generated** *by the $\{K_i : 1 \leq i \leq u\}$ is defined as follows:*

(a) $|W| = \bigcup_{i=1}^{u} |K_i|$.

(b) $e \in W$ if and only if e is equal to the closure of a maximal nonempty intersection of sets in the collection $\{e_i - \partial e_i, |W| - |K_i| : e_i \in K_i, 1 \leq i \leq u\}$.

REMARK 2.5.1. The partition W need not be a linear polyhedron in general. However, if the $\{K_i : 1 \leq i \leq u\}$ are in transverse position to one another (as in 2.6, 2.7, or 2.8), then W will be a full linear polyhedron.

Recall that a collection $\{S_i : i \in I\}$ of smooth submanifolds of the smooth manifold M are in **transverse position** to one another in M if the following hold:

(a) For every subset $J \subset I$ the intersection $\bigcap_{j \in J} S_j$ is a smooth submanifold of M.

(b) For any $i \in I - J$ and any $p \in S_i \bigcap (\bigcap_{j \in J} S_j)$ we have that $T(S_i)_p$ and $T(\bigcap_{j \in J} S_j)_p$ span $T(M)_p$.

DEFINITION 2.6. *A finite collection of linear polyhedra $\{K_i : 1 \leq i \leq u\}$ are in* **transverse position** *to one another in R^n if for every selection of members $e_i \in K_i$, $1 \leq i \leq u$, we have that the submanifolds $\{e_i - \partial e_i : 1 \leq i \leq u\}$ intersect with one another transversely in R^n.*

DEFINITION 2.7. *Two linear polyhedra K_1, K_2 in R^n are in* **transverse position** *to one another in R^n* **modulo** *a third linear polyhedron K_3 if the following hold:*

(a) There are subcomplexes $K_i' \subset K_i$, $i = 1,2$ such that $|K_1'| = |K_2'| = |K_3|$, and such that K_1' and K_2' both subdivide K_3.

(b) For any $e_1 \in K_1 - K_1'$ and $e_2 \in K_2 - K_2'$ we have that $(e_1 - \partial e_1)$ and $(e_2 - \partial e_2)$ are in transverse position in R^n.

(c) For any $e_1 \in K_1'$ and $e_2 \in K_2'$, such that $(e_i - \partial e_i) \subset (e_3 - \partial e_3)$ for some $e_3 \in K_3$ and $i = 1,2$, we have that $(e_1 - \partial e_1)$ and $(e_2 - \partial e_2)$ are in transverse position in $(e_3 - \partial e_3)$.

We now want to generalize 2.7 to make sense of a collection of linear polyhedra in R^n being in transverse position to one another "modulo" another collection of linear polyhedra in R^n. In the following definition let B denote a finite regular cell complex (recall that "regular" means that the boundary of each cell of B is embedded in B as a subcomplex). Let A denote a subcomplex of B, and let K_A denote a collection of full linear polyhedra in R^n—one such polyhedron K_e for each $e \in A$. For any subcomplex $C \subset A$ let $K(C)$ denote the partition generated by the $\{K_e : e \in C\}$.

DEFINITION 2.8. *The collection K_A of full linear polyhedra in R^n are said to be in B-***transverse position** *to one another if the following holds. For any subcomplex $C \subset A$, contained in a cell of B, $K(C)$ is a linear polyhedron. For any $e \in B$, and subcomplexes C_1, C_2 of $e \cap A$, we have that $K(C_1)$ and $K(C_2)$ are in transverse position in R^n modulo $K(C_1 \cap C_2)$.*

METRIC NOTATION 2.9. For any finite linear polyhedron K in R^n denote by $D(K)$ the minimum distance between any two disjoint members $e, e' \in K$. For each $e \in K$ denote by $d(e)$ the maximum distance between any two points of e. Set $\tau(K) = \min\{D(K)/d(e) : e \in K\}$. We call $\tau(K)$ the **thickness** of K. Set $\overline{D}(K) \equiv (\tau(K))^{-1}D(K)$. We call $\overline{D}(K)$ the **diameter** of K. For A, B, K_A as in 2.8 define $D(K_A)$ to be the minimum of all $\{D(K(e \cap A)) : e \in B\}$; and define $\tau(K_A)$ to be the minimum of all $\{D(K_A)/\overline{D}(K_e) : e \in A\}$. For any $e \in B$ let $N(e)$ denote the subcomplex of B consisting of all closed cells of B which intersect e. Let $N(B)$ denote the maximal number of cells which occur in any one of the $N(e)$, $e \in B$.

We can now state the first of the two propositions of this section.

PROPOSITION 2.10. *Let A, B, K_A be as in 2.8, and let r denote a given positive integer. Suppose $\tau(K_A) > \delta$, for some $\delta > 0$. Then the collection K_A extends to a collection K_B of full linear polyhedra in R^n which are in B-transverse position to one another in R^n and which also satisfy the following:*

(a) *There is a number $\delta' > 0$, which depends only on $(\delta, n, N(B), r)$ such that δ' is a lower bound for $\tau(K_B), D(K_B)/D(K_A)$.*

(b) *Given any compact subset $S \subset R^n$ it may be assumed that $S \subset |K_e|$ holds for each $e \in B - A$.*

(c) *For each $e \in B - A$ there is a linear triangulation L_e in R^n and an r-fold derived subdivision $L_e^{(r)}$ of L_e satisfying the following properties: $K_e = L_e^{(r)}$; there is a subcomplex L'_e of L_e such that L'_e subdivides $K(\partial e)$.*

Before stating the next proposition we need some more notation and another definition.

NOTATION 2.11. For any linear polyhedron K in R^n and any number $\varepsilon > 0$, denote by $K\#\varepsilon$ the maximal number of member sets in K which intersect any given ball of radius ε in R^n.

DEFINITION 2.12 *Let* $\{K_i : i \in I\}$ *denote a finite collection of linear polyhedra in* R^n, *and let* α *denote a positive number. The* $\{K_i : i \in I\}$ *are said to be in* α-**transverse position** to one another in R^n if for any $J \subset I$, such that J has more than n members, and for any collection $\{f_j : R^n \to R^n : j \in J\}$ of maps, satisfying $|p - f_j(p)| \leq \alpha$ for all $j \in J$ and all $p \in R^n$, we have that $\cap_{j \in J} f_j(|K_j^{n-1}|) = \emptyset$.

In the following proposition $K, \{K_{i,j} : 1 \leq i \leq n, j \in J_i\}$, and $\{K_i : 1 \leq i \leq b\}$ (where $n \leq b$) denote a finite number of full linear polyhedra in R^n, and $L^{(\beta)}$ is a β-fold derived subdivision of the linear triangulation L, all satisfying 2.13(a)-(g).

2.13. (a) K_i is generated by the $\{K_{i,j} : j \in J_i\}$ if $1 \leq i \leq n$.

(b) All of the polyhedra K, L, and $\{K_{i,j} : 1 \leq i \leq n, j \in J_i\}$ are in transverse position to one another.

(c) All of the polyhedra K, $\{K_{i,j} : 1 \leq i \leq n, \ j \in J_i\}$ are in α-transverse position to one another for some $\alpha > 0$.

(d) $\overline{D}(L) < \alpha/2$; and $K\#\overline{D}(L) + L\#\overline{D}(L) + \sum_{i=1}^{n} K_i\#\overline{D}(L) < \gamma$ for some $\gamma > 0$.

(e) Let $\delta > 0$ be a lower bound for any of the $\tau(X)$, $D(X)/D(Y)$, where X, Y may stand for any of $L^{(\beta)}$, $\{K_i : n+1 \leq i \leq b\}$, $\{V\}$. (Here $\{V\}$ is the collection of all linear polyhedra generated by any subcollection of the $\{K_{i,j} : 1 \leq i \leq n, j \in J_i\} \cup \{K\} \cup \{L\}$.)

(f) The maximal number of the $\{(K_{i,j}) : 1 \leq i \leq n, j \in J_i\}$ which intersect any given $|K_{i',j'}|$ (with $1 \leq i' \leq n, j' \in J_{i'}$) is less than γ.

(g) $|K| \cup (\cup_{i=1}^{n} |K_i|) \subset \mathrm{Int}(|L|)$.

PROPOSITION 2.14. *There is a positive number* β, *which depends only on* (n, γ), *such that for any given* β*-fold derived subdivision* $L^{(\beta)}$ *of* L *there are homeomorphisms*

$$f_i : R^n \to R^n,$$

for $i = 1, 2, \ldots, n+1$ *which satisfy the following properties*

(a) $|f_i(p) - p| < \overline{D}(L)$ *for any* $p \in R^n$ *and any* $i \in 1, 2, \ldots, n+1$. $f_i(e) = f_{i+1}(e)$ *for any* $i \in \{1, 2, \ldots, n\}$ *and any* $e \in K_i$. *Let* L' *be the smallest subcomplex of* L *such that* $|K| \subset \mathrm{Int}(|L'|)$. *Then* $f_i(p) = p$ *for all* $p \in R^n - |L'|$ *and all* $i \in \{1, 2, \ldots, n\}$. *Moreover* $f_{n+1} = 1$.

(b) $f_1(e)$ *is a subcomplex of* $L^{(\beta)}$ *for each* $e \in K$. *There is a full linear polyhedron* $K'_{i,j}$ *(for each* $j \in J_i$ *and each* $i \in \{1, 2, \ldots, n\}$*) such that for each* $e \in K_{i,j}$ *we have that* $f_i(e)$ *is a subcomplex of* $K'_{i,j}$. *Moreover* $|K'_{i,j}| = f_i(|K_{i,j}|)$, *and* $K'_{n,j} = K_{n,j}$ *for all* $j \in J_n$.

(c) For each $k \in \{n+1, n+2, \ldots, b\}$ *all the polyhedra* K_k, $\{K'_{i,j} : 1 \leq i \leq n-1,\ j \in J_i\}$ *are in transverse position to one another. The* $\{K'_{i,j} : 1 \leq i \leq n,\ j \in J_i\}$ *are in* α'*-transverse position to one another, where* $\alpha' = \alpha/4$. *Moreover* $\sum_{i=1}^{n} K'_i \# \overline{D}(L) < \gamma'$, *where* $\gamma' > 0$ *depends only on* (n, γ). *Here* K'_i *is the linear polyhedron generated by the* $\{K'_{i,j} : j \in J_i\}$. *All the polyhedra* $\{K'_{i,j} : 1 \leq i \leq n,\ j \in J_i\}$ *are in transverse position to one another.*

(d) There is $\delta' > 0$, *which depends only on* (δ, n, b, γ), *such that* δ' *is a lower bound for all the* $\tau(X')$, $D(X')/D(Y')$, *where* X', Y' *may stand for any linear polyhedron generated by any subcollection of any collection* $\{K'_{i,j} : 1 \leq i \leq n-1,\ j \in J_i\} \cup \{K_k\}$ *(here we have a different collection for each* $k \in \{n+1, n+2, \ldots, b\}$*), or by any subcollection of* $\{K'_{i,j} : 1 \leq i \leq n,\ j \in J_i\}$.

§3. PROOFS OF PROPOSITIONS 2.10 AND 2.14

We first state five lemmas which are used in the proofs of propositions 2.10 and 2.14. We then use these lemmas to prove 2.10 and 2.14. Finally at the end of this section the five lemmas are proven.

LEMMA 3.1. *Let K be a full linear polyhedron in R^n, and let T denote a full linear triangulation in R^n. Suppose $\delta > 0$ is a lower bound for all of $\tau(K), \tau(T), (D(K)/D(T))^{\pm 1}$. Then for any given $\varepsilon \in (0,1)$ the following are true:*

(a) There is a P. L. flow $\varphi_t : T \to T_t$, $t \in [0,1]$, of T such that each vertex is moved a distance no more than $\varepsilon D(T)$ from its initial position and such that K and T_1 are in transverse position.

(b) There is $\delta' > 0$, depending only on (δ, ε, n), such that for all $0 \leq t \leq 1$ δ' is a lower bound for all of $\tau(T_t)$, $\tau(W)$, $(D(T)/D(T_t))^{\pm 1}$, $(D(T)/D(W))^{\pm 1}$. Here W is the linear polyhedron generated by K and T_1.

(c) Suppose that K and T are already in transverse position and let $\beta > 0$ be a lower bound for all of $\tau(K), \tau(T), \tau(V), D(V)/D(K)$, $D(V)/D(T)$, where V is the polyhedron generated by K and T. Then there is $\gamma > 0$, which depends only on (β, n), such that for any P. L. flow $\varphi_t : T \to T_t, t \in [0,1]$, which moves each vertex of T a distance less than $\gamma D(T)$ we have that K and T_1 are still in transverse position. Furthermore there is $\beta' > 0$, which depends only on (β, n), such that β' is a lower bound for all $\tau(V_1), (D(V)/D(V_1))^{\pm 1}$, where V_1 is the linear polyhedron generated by K and T_1.

LEMMA 3.2. *Let K, L denote full linear polyhedra in R^n which are in transverse position to one another modulo a third linear polyhedron M. Let T, S denote full linear triangulations satisfying the following properties: S is a subcomplex of T; S is a subdivision of L; for any subcomplex L' of L and $\Delta \in T$ we must have that $\Delta \cap |L'|$ is a simplex. Suppose $\delta > 0$ is a lower bound for $\tau(L), \tau(T), D(T)/D(L), \tau(W), (D(W)/D(T))^{\pm 1}$, $\tau(K), (D(K)/D(T))^{\pm 1}$, where W is the linear polyhedron generated by K and L. Then for any given $\varepsilon \in (0,1)$ the following are true.*

(a) There is a P. L. flow $\varphi_t : T \to T_t, t \in [0,1]$, which moves each vertex of T a distance less than $\varepsilon D(T)$, such that $\varphi_t(e) = e$ holds for every $e \in L, t \in [0,1]$. Moreover T_1 is in transverse position to K modulo M.

(b) There is $\delta' > 0$, which depends only on (δ, ε, n), such that for all $0 \leq t \leq 1$ δ' is a lower bound for all of $\tau(T_t), \tau(V), (D(T_t)/D(T))^{\pm 1}$,

and $D(V)/D(T_1)$. Here V is the linear polyhedron generated by K and T_1.

LEMMA 3.3. *Let K, L, M, T, S, be as in 3.2. Suppose that T is in transverse position to K modulo M. We let V denote the linear polyhedron generated by K and T. Let $\delta_1 > 0$ be a lower bound for all $\tau(V)$ and $D(V)/D(T)$. Then there are numbers $\delta_1', \varepsilon > 0$ which satisfy the following properties.*

(a) δ_1', ε depend only on (δ, δ_1, n), where δ comes from 3.2.

(b) Suppose $\varphi_t : T \to T_t, t \in [0,1]$, is a P. L. flow of T such that no vertex of T is moved a distance greater than $\varepsilon \cdot D(T)$, and such that for each $e \in L$ and each $t \in [0,1]$ we have $\varphi_t(e) = e$. Then each T_t intersects with K transversely modulo M, and for each $0 \leq t \leq 1$, δ_1' is a lower bound for all $\tau(V_t)$, $\tau(T_t)$, $(D(T_t)/D(T))^{\pm 1}$, $(D(V_t)/D(T))^{\pm 1}$. Here V_t is the linear polyhedron generated by K and T_t.

LEMMA 3.4. *Let K denote a full linear polyhedron in R^n; let L denote a full linear triangulation in R^n which is in transverse position to K; let W denote the linear polyhedron generated by K and L. Suppose $\delta > 0$ is a lower bound for all of $\tau(W), \tau(L), D(W)/D(L)$. Suppose also that for $\gamma > 0$ we have $K \# \overline{D}(L) \leq \gamma$. Then there is a positive integer r and an r-fold derived subdivision $L^{(r)}$ of L which satisfy the following:*

(a) For each $e \in K$ and $e' \in L$ the set $e \cap e'$ is a subcomplex of $L^{(r)}$.

(b) The lower bound for both $\tau(L^{(r)})$ and $D(L^{(r)})/D(L)$ depends only on (δ, n).

(c) An upper bound for r depends only on (γ, n).

(d) For any $e' \in L^{(r)}$, $e \in W$ with $e' \subset e$, we must have that $e' \cap \partial e$ is a simplex.

REMARK 3.5. Suppose in lemma 3.4 an upper bound for $K \# \overline{D}(L)$ is not given. We can always choose an upper bound for $K \# \overline{D}(L)$ which is dependent only on (δ, n). Then by applying lemma 3.4 we can obtain $L^{(r)}$ satisfying 3.4(a),(b),(d), and satisfying (instead of 3.4(c)) the following:

(c′) An upper bound for r depends only on (δ, n).

In the following lemma let $K, \{L_i : 1 \leq i \leq u\}$ be linear polyhedra in R^n which are in transverse position to one another. We assume further that K is a linear triangulation and $\varphi_t : R^n \to R^n, t \in [0,1]$, is a P. L. isotopy which satisfy the following properties.

3.6. (a) $\varphi_t(x) = x$ for all $x \in R^n - |K|$ and all $t \in [0,1]$.

(b) The restriction $\varphi_t \| K|$ yields a P. L. flow $\varphi_t : K \to K_t$ of the linear triangulation K (see 2.3).

(c) Note each $\varphi_t(|L_i|)$ has the structure of a linear polyhedron $L_{i,t}$ defined as follows. Let W_i denote the linear polyhedron generated by L_i

and K. Let $L_{i,o}^{n-1}$ denote the subcomplex of W_i which subdivides L_i^{n-1}. Set $L_{i,o} - L_{i,o}^{n-1} = L_i - L_i^{n-1}$. Finally set $L_{i,t}$ equal the image of $L_{i,o}$ under $\varphi_t : R^n \to R^n$.

Also in the following lemma T_i will denote a linear triangulation in R^n, and $\varphi_{i,t} : R^n \to R^n, t \in [0,1]$, denote P.L. isotopies satisfying the following.

(d) T_i is a subdivision of $L_{i,1}$.

(e) $\varphi_{i,t}(x) = x$ for all $x \in R^n - |T_i|$ and all $t \in [0,1]$; the restriction of $\varphi_{i,t} : R^n \to R^n, t \in [0,1]$, yields a P.L. flow of T_i.

LEMMA 3.7. *Let $\delta > 0$ be a lower bound for all of $\tau(T_i), \tau(V)$, $(D(V)/D(T_i))^{\pm 1}$, where V is the linear polyhedron generated by all the $K_1, \{L_{i,1} : 1 \leq i \leq u\}$. Then there is $\varepsilon > 0$, which depends only on (δ, n, u), such that if $|\varphi_{i,t}(x) - x| < \varepsilon D(T_i)$ holds for all $x \in R^n$, all $i \in \{1, 2, \ldots, u\}$, and all $t \in [0,1]$, then there is a homeomorphism $g : R^n \to R^n$ satisfying the following.*

(a) $g(x) = x$ for all $x \in R^n - \bigcup_{i=1}^u |L_{i,1}| - |K|$.

(b) $g(e) = \varphi_1(e)$ for all $e \in K$; $g(e) = \varphi_{i,1} \circ \varphi_1(e)$ for all $e \in L_i$ and all $i \in \{1, 2, \ldots, u\}$.

REMARK 3.7$'$. There is a stronger version of Lemma 3.7 in which ε depends only on δ, n, u', where u' = maximum number of the $\{L_i : 1 \leq i \leq u\}$ which intersect any given L_k.

PROOF OF PROPOSITION 2.10: For each $i \in \{0, 1, \ldots, \dim(B)\}$ there is a positive integer i' and a collection of subsets $\{J_{i,j} : 1 \leq j \leq i'\}$ of $B^i - (A \cup B^{i-1})$ which satisfy the following:

3.8. (a) $B^i - (A \cup B^{i-1}) = \bigcup_{j=1}^{i'} J_{i,j}$; $J_{i,j} \bigcap J_{i,j'} = \emptyset$ if $j \neq j'$.

(b) If for some $e \in B$ and for some (i,j) and $e', e'' \in J_{i,j}$ we have $e \cap e' \neq \emptyset$ and $e \cap e'' \neq \emptyset$, then we must have $e' = e''$.

(c) There is an upper bound for i' which depends only on $(i, N(B))$.

Set $B(i,j) = A \cup B^{i-1} \cup \left(\cup_{k=1}^{j} J_{i,k}\right)$. The proof of 2.10 will proceed by induction over the sequence $\ldots, B(i,j), B(i,j+1), \ldots$. Here is the induction hypothesis.

3.9(i,j). The collection K_A extends to a collection $K_{B(i,j)}$ of full linear polyhedra in R^n which are in B-transverse position to one another. Moreover 2.10(b)(c) hold for each $e \in B(i,j)$, and there is a lower bound for $\tau(K_{B(i,j)})$ and $D(K_{B(i,j)})/D(K_A)$ which depends only on $(\delta, n, \alpha_{i,j}, r)$, where $\alpha_{i,j} = \left(\sum_{k=0}^{i} k'\right) + j$.

Now we must show that 3.9(i,j) $\Rightarrow$ 3.9(i,j+1) (or show that 3.9(i,j) $\Rightarrow$ 3.9$(i+1,1)$, if $j = i'$).

Use lemma 3.1 to choose for each $e \in B(i,j+1) - B(i,j)$ a linear triangulation T_e which is in transverse position to $K(\partial e)$ and such that there is a lower bound for all $\tau(W_e), D(W_e)/D(T_e)$, $D(W_e)/D(K(\partial e))$, which depends only on n and the lower bound for 3.9(i,j). (Here W_e denotes the linear polyhedron generated by T_e and $K(\partial e)$.) We may also suppose that $|T_e| \supset |K(\partial e)| \cup S$, where S comes from 2.10(b).

Use lemma 3.4, 3.5 as applied to $K(\partial e)$ and T_e, to get a linear triangulation L_e satisfying:

<u>3.10.</u> (a) L_e subdivides T_e. There is a full subcomplex of L_e which subdivides $K(\partial e)$.

(b) There is a lower bound for both $\tau(L_e)$ and $D(L_e)/D(K(\partial e))$ which depends only on n and the lower bound in 3.9(i,j).

Now we list, for each $e \in B(i,j+1) - B(i,j)$, the subcomplexes $X_1, X_2, \ldots, X_{\beta(e)}$ of $B(i,j)$ which satisfy $X_i \cup e \subset f$ for some $f \in B$. By arguing inductively over the sequence $X_1, X_2, \ldots$, using alternately lemma 3.2 and lemma 3.3, we can move L_e through a P. L. flow (which leaves each set in $K(\partial e)$ invariant) to a new linear triangulation—also denoted by L_e—so that the new L_e will satisfy the following in addition to 3.10(a)(b).

<u>3.11.</u> L_e is in transverse position to each $K(X_k)$ modulo $K(X_k \cap e)$. Let $V_{e,k}$ denote the linear polyhedron generated by $K(X_k)$ and L_e. Then there is a lower bound for all the $\tau(V_{k,e})$ and $D(V_{k,e})/D(L_e)$ which depends only on $n, \beta(e)$, and the lower bound in 3.9(i,j).

Next choose a sequence $L_e^{(1)}, L_e^{(2)}, \ldots, L_e^{(r)}$ of higher order derived subdivisions of L_e which satisfy the following:

<u>3.12.</u> (a) There is a lower bound for $\tau(L_e^{(k)})$ and $D(L_e^{(k)})/D(L_e), 1 \leq k \leq r$, which depends only on (r,n) and the lower bound in 3.11.

We now argue inductively over the sequence $X_1, X_2, \ldots$ of the preceding paragraph, using lemma 3.2 and lemma 3.3 alternately, to get a P. L. flow of $L_e^{(r)}$ which leaves each triangle of L_e invariant and moves $L_e^{(r)}$ to another r-fold derived subdivision of L_e—also denoted by $L_e^{(r)}$—which satisfies the following in addition to 3.12(a).

<u>3.12.</u> (b) $L_e^{(r)}$ is in transverse position to each $K(X_k)$ modulo $K(X_k \cap e)$. Let $V'_{e,k}$ denote the linear polyhedron generated by $K(X_k)$ and $L_e^{(r)}$.

(c) There is a lower bound for all $\tau(V'_{e,k})$ and $D(V'_{e,k})/D(L_e)$ which depends only on (n, r) and the lower bounds of 3.11 and 3.12(a).

Finally define, for each $e \in B(i, j+1) - B(i, j)$, K_e to be the linear polyhedron $L_e^{(r)}$. This completes the extension of $K_{B(i,j)}$ to $K_{B(i,j+1)}$. It is left as an exercise to deduce from 3.10-3.12 that $K_{B(i,j+1)}$ satisfies 3.9(i,j+1).

This completes the proof of proposition 2.10.

PROOF OF PROPOSITION 2.14.

First we establish the integer β of 2.14. We denote by W_1 the linear polyhedron generated by K and L, by W_2 that generated by L and all the $\{K_i : 1 \leq i \leq n\}$, and by W_3 that generated by W_1 and W_2. Since $K\#\overline{D}(L) \leq \gamma$, we may apply lemma 3.4 to K and L to get a subdivision L' of L satisfying:

3.13. (a) L' is β-fold derived subdivision of L. L' contains a subcomplex which subdivides K.

(b) β depends only on (γ, n).

(c) Given a lower bound $\sigma > 0$ for $\tau(W_1), \tau(L), D(W_1)/D(L)$, $D(W_1)/D(K)$ there is a lower bound for $\tau(L')$ and $D(L')/D(L)$ which depends only on (σ, n).

(d) For any $e' \in L', e \in W_1$, if $e' \subset e$ then $e' \cap \partial e$ is a simplex.

Note that W_1 and W_2 are in transverse position modulo L, and L' is a subdivision of W_1. It follows that lemma 3.2 may be applied to L' to get a P. L. flow of L' (which leaves each $e \in W_1$ invariant) that flows L' to another β-fold derived subdivision of L—also denoted by L'—which satisfies the following in addition to 3.13(a)(b)(c).

3.13. (e) L' and W_2 are in transverse position to one another modulo L.

(f) Given a lower bound $\sigma' > 0$ for $\tau(L), \tau(W_3)$, and $(D(W_3)/D(L))$, there is a lower bound for $\tau(V)$ and $D(V)/D(L')$ depending only on σ, σ', n. Here V is generated by L' and W_2.

Now let $L^{(\beta)}$ denote any given β-fold derived subdivision of L. Define subcomplexes $N_i, i = 1, 2, \ldots, n+1$, of L as follows.

3.14. (a) N_1 is the union of all closed triangles $e \in L$ such that $K^{n-1} \cap e \neq \emptyset$.

(b) N_{i+1} is the union of all $e \in N_i$ such that $K_i^{n-1} \cap e \neq \emptyset$.

For each $i \in \{1, 2, \ldots, n+1\}$ define a β-fold derived subdivision L'_i of L as follows.

3.15. L'_i has the same vertices as does L' (of 3.13) in the subcomplex $\overline{L - N_i}$; L'_i has the same vertices as does $L^{(\beta)}$ in $\text{int}(N_i)$.

We can now define the $f_i : R^n \to R^n$ of 2.14.

<u>3.16.</u> Define $f_i : R^n \to R^n$, for each $1 \leq i \leq n+1$, to be the canonical simplicial isomorphism $L' \to L'_i$ between the two β-fold derived subdivisions L' and L'_i of L. (This defines $f_i \big| |L|$; set $f_i \big| (R^n - |L|) =$ identity.)

Towards verifying 2.14(a) note that $|f_i(p) - p| \leq \overline{D}(L)$ for all $p \in |L|$ because $f_i : R^n \to R^n$ leaves invariant each $e \in L$, and $f_i(p) = p$ for all $p \in R^n - |L|$. Note that $N_{n+1} = \emptyset$ because the $\{K, K_1, K_2, \dots, K_n\}$ are in α-transverse position to one another and $\overline{D}(L) \leq \alpha/2$. From $N_{n+1} = \emptyset$ and 3.15 and 3.16 it follows that $f_{n+1} = 1$. Finally $f_i(e) = f_{i+1}(e)$ for any $e \in K_i$ follows from 3.14(b), 3.15, 3.16.

To verify 2.14(b) note that $f_1(e)$ is a subcomplex of $L^{(\beta)}$ by 3.13(a), 3.14(a), 3.15, 3.16. For each $i \in \{1, 2, \dots, n-1\}$ and $j \in J_i$ let $V_{i,j}$ denote the linear polyhedra generated by $K_{i,j}$ and L'. Note that the image of $V_{i,j}$ under $f_i : R^n \to R^n$ is a linear polyhedron $V'_{i,j}$ in R^n. For $i < n$ construct $K'_{i,j}$ from $V'_{i,j}$ as follows: if $e \in V'_{i,j}$ and $e \cap f_i(e' - \partial e') = \emptyset$ for any $e' \in K_{i,j} - K^{n-1}_{i,j}$ then $e \in K'_{i,j}$; $f_i(e') \in K'_{i,j}$ for any $e' \in K_{i,j} - K^{n-1}_{i,j}$. Set $K'_{n,j} = K_{n,j}$ for $j \in J_n$.

Towards verifying 2.14(c) we first note that the collection $\{K'_{i,j} : 1 \leq i \leq n, j \in J_i\}$ are in $\alpha/2$-transverse position to one another for the following reasons: the $\{K_{i,j} : 1 \leq i \leq n, j \in J_i\}$ are in α-transverse position; $f_i(|K^{n-1}_{i,j}|) = |(K'_{i,j})^{n-1}|$ for all $i \in \{1, 2, \dots, n\}$, $j \in J_i$; $|f_i(p) - p| \leq \overline{D}(L)$ for all $p \in R^n$ and all i; $\overline{D}(L) \leq \alpha/2$. Next note it follows from 3.13(b), and the hypothesis 2.13(d) for 2.14, that there is $\gamma_1 > 0$ depending only on (n, γ) such that $\sum_{i=1}^n K'_i \# \overline{D}(L) < \gamma_1$. Unfortunately at this point the $\{K'_{i,j} : 1 \leq i \leq n, j \in J_i\}$ need not be in transverse position to one another or to any of the $\{K_i : n+1 \leq i \leq b\}$. To remedy this we use lemmas 3.1, 3.4 to choose a linear triangulation $\{T_{i,j} : 1 \leq i \leq n-1, j \in J_i\}$ of the $\{V'_{i,j} : 1 \leq i \leq n-1, j \in J_i\}$ and we use lemma 3.1 to choose for each $T_{i,j}$ a P.L. flow $\varphi_{i,j,t} : T_{i,j} \to T_{i,j,t}$, $t \in [0,1]$, such that the following properties hold (compare with 2.13(d)(f)).

<u>3.17.</u> (a) There is $\gamma_2 > 0$, which depends only on (γ, n), such that $\sum_{i,j} T_{i,j,t} \# \overline{D}(L) < \gamma_2$ holds for all $t \in [0,1]$, where the summation runs over all $i \in \{1, 2, \dots, n\}$, and all $j \in J_i$.

(b) For each $k \in \{n+1, \dots, b\}$ all the linear polyhedra K_k, $\{T_{i,j,1} : 1 \leq i \leq n-1, j \in J_i\}$ are in transverse position to one another. Moreover all the polyhedra $\{T_{i,j,1} : 1 \leq i \leq n-1, j \in J_i\} \cup \{K'_{n,j} : j \in J_n\}$ are in transverse position to one another.

Now define a new set of $\{K'_{i,j} : 1 \leq i \leq n-1, j \in J_i\}$ by using the $\{T_{i,j,1} : 1 \leq i \leq n-1, j \in J_i\}$ in place of the $\{V'_{i,j} : 1 \leq i \leq n-1, j \in$

$J_i\}$ and proceeding to define the new $\{K'_{i,j}\}$ from the $\{T_{i,j,1}\}$ exactly as the old $\{K'_{i,j}\}$ were defined from the $\{V'_{i,j}\}$. That is if $e \in T_{i,j,1}$ and $e \cap \varphi_{i,j,1} \circ f_i(e' - \partial e') = \emptyset$ holds for all $e' \in K_{i,j} - K^{n-1}_{i,j}$, then $e \in K'_{i,j}$; $\varphi_{i,j,1} \circ f_i(e') \in K'_{i,j}$ for any $e' \in K_{i,j} - K^{n-1}_{i,j}$. Note if the P.L. flows $\varphi_{i,j,t} : T_{i,j} \to T_{i,j,t}$, $t \in [0,1]$, are chosen sufficiently close to the identity map than the new $\{K'_{i,j} : 1 \le i \le n, j \in J_i\}$ will still be in $\alpha/4$-transverse position to one another. This last remark and 3.17 imply that the new $\{K'_{i,j} : 1 \le i \le n, j \in J_i\}$ satisfy 2.14(c).

On the other hand the old f_i do not satisfy 2.14(b) for the new $\{K'_{i,j}\}$. So we now need to define new $\{f_i\}$ which will satisfy both 2.14(a)(b) for the new $\{K'_{i,j}\}$. Towards this end we note that (by lemma 3.7) if the P.L. flows $\varphi_{i,j,t} : T_{i,j} \to T_{i,j,t}$, $t \in [0,1]$, are sufficiently close to the identity map then there are homeomorphisms $g_i : R^n \to R^n$ for each $i \in \{1, 2, \dots, n\}$ satisfying the following.

3.18. (a) $g_1(e) = f_1(e)$ for all $e \in K; g_1(e) = \varphi_{1,j,1} \circ f_1(e)$ for all $j \in J_1$ and all $e \in K_{1,j}$.

(b) $g_i(e) = \varphi_{i-1,j,1} \circ f_i(e)$ for all $i \in \{2, 3, \dots, n\}$, $j \in J_{i-1}$, and $e \in K_{i-1,j}$; $g_i(e) = \varphi_{i,j,1} \circ f_i(e)$ for all $i \in \{2, 3, \dots, n\}$, $j \in J_i$, and $e \in K_{i,j}$, where $\varphi_{n,j,t}(x) = x$ for all $j \in J_n, t \in [0,1]$, $x \in R^n$.

(c) $g_i(x) = x$ for all $i \in \{1, 2, \dots, n\}$ and all $x \in R^n - |L|$.

Now we can define the new homeomorphisms $f_i : R^n \to R^n$ as follows.

3.19. new $f_i = g_i$ if $i \in \{1, 2, \dots, n\}$;
new $f_{n+1} = 1$.

Note it follows from 3.18, the description of the new $K'_{i,j}$, and the fact that the old $\{f_i\}$ satisfied 2.14(a), that the new $\{K'_{i,j}\}$ and the new $\{f_i\}$ satisfy 2.14(a)(b).

It is left as an exercise to show that the P.L. flows $\varphi_{i,j,t} : T_{i,j} \to T_{i,j,t}$, $t \in [0,1]$, of 3.17 can be chosen so that the new $K'_{i,j}$ and the new $f_i : R^n \to R^n$ satisfy 2.14(d) as well as 2.14(a)(b)(c).

This completes the proof of proposition 2.14.

The proofs of lemmas 3.1, 3.2, 3.3, 3.4, 3.7 will now be given.

PROOF OF LEMMA 3.1: We say that the intersection of the two linear polyhedra K, T (with T a linear triangulation) is β-stable (for some $\beta > 0$) if any P.L. flow $\varphi_t : T \to T_t$, $t \in [0,1]$, which satisfies (a) below also must satisfy (b) below.

(a) No vertex of T is moved a distance greater than β by the flow $\varphi_t : T \to T_t$, $t \in [0,1]$.

(b) For any $e \in K$, $e' \in T$, and $t \in [0,1]$, we have that $e \cap \varphi_t(e') \neq \emptyset$ if and only if $e \cap e' \neq \emptyset$.

The verification of the following claim is left as an exercise.

Claim 3.20. (a) K and T are in transverse position to one another if and only if their intersection is β-stable for some $\beta > 0$.

(b) Suppose that $\delta > 0$ is a lower bound for each of $\tau(K), \tau(T)$, $(D(K)/D(T))^{\pm 1}$ and that the intersection of K, L is $\gamma D(T)$-stable for some $\gamma > 0$. Then there is $\delta' > 0$, which depends only on (δ, γ, n), such that δ' is a lower bound for each of $\tau(V), (D(V)/D(T)$. Here V is the linear polyhedron generated by K and T.

(c) Suppose that K and T are in transverse position and that $\delta > o$ is a lower bound for all of $\tau(K), \tau(T), \tau(V), D(V)/D(T), D(V)/D(K)$. Then there is $\gamma > 0$, which depends only on (δ, n), such that the intersection of K and T is $\gamma D(T)$-stable.

Note that 3.1(c) can be deduced directly from 3.20.

Note also, as a consequence of 3.20, that a P.L. flow $\varphi_t : T \to T_t$, $t \in [0,1]$, will satisfy 3.1(a)(b) if it satisfies the following properties.

3.21. (a) The flow $\varphi_t : T \to T_t$ moves no vertex of T a distance greater than $\varepsilon \cdot D(T)$.

(b) There is $\gamma > 0$, which depends only on (ε, δ, n), such that the intersection of K and T_1 is $\gamma D(T)$-stable.

(c) γ is also a lower bound for all $\tau(T_t), (D(T)/D(T_t))^{\pm 1}$.

We use an induction argument to verify 3.21. Before stating our induction hypothesis we need some notation.

For each $e \in T$ let $S(e, K)$ denote the subcomplex of all $e' \in K$ such that distance$(e, e') \leq D(T)$. Note there is $\beta > 0$ satisfying the following properties.

3.22. (a) β depends only on (δ, n).

(b) The number of simplices in each $S(e, K)$ is less than β.

(c) For each $r \in \{-1, 0, 1, \ldots, n-1\}$ the collection of $(r+1)$-triangles $T^{r+1} - T^r$ is equal a disjoint union $\cup_{i=1}^{\beta} J_{i,r}$. If $e_1, e_2 \in J_{i,r}$ for some $i \in \{1, 2, \ldots, \beta\}$ then $e_1 \cap e_2 = \emptyset$.

For each $r \in \{0, 1, \ldots, n\}$ and each $j \in \{1, 2, \ldots, \beta\}$ set $T^{r,j} = T^r \cup (\cup_{e \in J_{1,r}} e) \cup (\cup_{e \in J_{2,r}} e) \cup \cdots \cup (\cup_{e \in J_{j,r}} e)$. Our induction argument is carried out over the sequence $\ldots \subset T^{r-1,\beta-1} \subset T^r \subset T^{r,1} \subset T^{r,2} \ldots \subset T^{r,\beta-1} \subset T^{r+1} \subset T^{r+1,1} \subset \ldots$. Here is our induction hypothesis.

Hypothesis 3.23(r, j). There is a P.L. flow $\varphi_{r,j,t} : T \to T_t$, $t \in [0,1]$, which satisfies the following properties.

(a) $\varphi_{r,j,t} : T \to T_t$, $t \in [0,1]$, moves no vertex of T a distance greater than $\varepsilon_{r,j} D(T)$, where $\varepsilon_{r,j} = ((r\beta + j)/(n\beta + \beta))\varepsilon$.

(b) There is $\gamma_{r,j} > 0$, which depends only on the $(\delta, \varepsilon, \beta, r, j, n)$, such that the intersection of K and $(T^{r,j})_1$ is $\gamma_{r,j} D(T)$-stable.

(c) $\gamma_{r,j}$ is also a lower bound for all $\tau(T_t), (D(T)/D(T_t))^{\pm 1}$.

The induction step consists of showing $3.23(r, j) \Rightarrow 3.23(r, j+1)$ if $j < \beta$ (or of showing $3.23(r, j) \Rightarrow 3.23(r+1, 1)$ if $j = \beta$).

Let $e \in J_{j+1,r}$; let $e_1, e_2, e_3, \ldots, e_\beta$ denote the simplices of $S(e, K)^{n-r-2}$ (see 3.22(b)); let P_i be the plane spanned by e_i. Set $e' = \varphi_{r,j,1}(e)$. We can choose a P.L. flow $\varphi_{e,t} : e' \to e'_t$, $t \in [0,1]$, to satisfy the following properties:

3.24. (a) $\varphi_{e,t} : e' \to e'_t$, $t \in [0,1]$, moves no vertex of e' a distance greater than the minimum $\left\{\frac{1}{2}\gamma_{r,j} D(T), (\varepsilon/(n+1)\beta) D(T)\right\}$.

(b) There is a number $\eta > 0$, which depends only on the $(\gamma_{r,j}, \beta, n, \varepsilon)$, such that for all i the distance from e'_1 to P_i is greater than $\eta D(T)$.

Towards verifying 3.24 we note that $\dim(P_i) + \dim(e') < n$, so by varying the positions of the vertices of e' a little bit (this variation is the P.L. flow $\varphi_{e,t} : e' \to e'_t$) we can move e' away from all the P_i. The remaining details in the verification of 3.24 are left as an exercise.

We can now define the P.L. flow $\varphi_{r,j+1,t} : T \to T_t$, $t \in [0,1]$, as follows.

3.25. For each vertex $v \in T$ set

$$\varphi_{r,j+1,t}(v) = \begin{cases} \varphi_{r,j,2t}(v), & \text{for } t \in [0, 1/2]; \\ \varphi_{r,j,1}(v), & \text{for } t \in [1/2, 1] \text{ and } v \notin \cup_{e \in J_{j+1,r}} e; \\ \varphi_{e,t} \circ \varphi_{r,j,1}(v), & \text{for } t \in [1/2, 1] \text{ and} \\ & \quad v \in e \text{ for some } e \in J_{j+1,r}. \end{cases}$$

It follows from 3.22(c) that if η and each $\varphi_t : e' \to e'_t$, $t \in [0,1]$, of 3.24 are chosen sufficiently close to 0 and the identity map, then $\varphi_{r,j+1,t} : T \to T_t$, $t \in [0,1]$, given in 3.25 is well defined and satisfies $3.23(r, j+1)$ (c) for $\gamma_{r,j+1} \le \frac{1}{2}\gamma_{r,j}$. Now it follows from this last remark and $3.23(r, j)$, 3.24, 3.25, that $\varphi_{r,j+1,t} : T \to T_t$, $t \in [0,1]$, satisfies $3.23(r, j+1)$ for

$$\gamma_{r,j+1} = \min\left\{\frac{1}{4}\eta, \frac{1}{4}\gamma_{r,j}\right\}.$$

This completes the induction step.

Finally note that $3.23(n, 1)$ implies 3.21 for $\gamma = \gamma_{n,1}$ in 3.21.

This completes the proof of lemma 3.1.

PROOF OF LEMMA 3.2, 3.3: We carry out the proof in the special case when $T = S$ in 3.2. This proof uses the same ideas as does the proof of 3.1, so we will only outline them.

Let K, L, T be as in 3.2, 3.3. We shall say that the intersection of K and T is **β-stable modulo L** (for some $\beta > 0$) if any P.L. flow $\varphi_t : T \to T_t,\ t \in [0,1]$, which satisfies (a) below must also satisfy (b).

(a) No vertex of T is moved a distance greater than β by the flow $\varphi_t : T \to T_t,\ t \in [0,1]$; for any $e \in L$ and any $t \in [0,1]$ we have $\varphi_t(e) = e$.

(b) For any $e \in K,\ e' \in T$, and any $t \in [0,1]$, we have that $e \cap \varphi_t(e') \neq \emptyset$ if and only if $e \cap e' \neq \emptyset$.

The verification of the following claim is left as an exercise. (In verifying 3.26 the hypothesis of 3.2, 3.3—that $e' \cap \partial e$ is a simplex whenever $e \in L,\ e' \in T,\ e' \subset e$—is needed.)

Claim 3.26. (a) K and T are in transverse position to one another modulo M if and only if their intersection is β-stable modulo L for some $\beta > 0$ (for M as in 3.2).

(b) Suppose that the intersection of K and T is $\gamma D(T)$-stable modulo L for some $\gamma > 0$. Then there is $\beta > 0$, which depends only on (γ, δ, n), such that β is a lower bound for each of $\tau(V)$ and $D(V)/D(T)$. (Here V is the linear polyhedron generated by K and T, and δ comes from 3.2.)

(c) Suppose that K and T are in transverse position modulo M and that $\beta > 0$ is a lower bound for $\tau(V)$ and $D(V)/D(T)$. Then there is $\gamma > 0$, which depends only on (β, δ, n), such that the intersection of K and T is $\gamma D(T)$-stable modulo L.

Note that lemma 3.3 can be deduced directly from 3.26.

Note also, as a consequence of 3.26, that a P.L. flow $\varphi_t : T \to T_t,\ t \in [0,1]$, will satisfy 3.2(a)(b) if it satisfies the following properties.

3.27. (a) The flow $\varphi_t : T \to T_t, t \in [0,1]$, moves no vertex of T a distance greater than $\varepsilon \cdot D(T)$; for any $e \in L$ and any $t \in [0,1]$ we have that $\varphi_t(e) = e$.

(b) There is $\gamma > 0$, which depends only on the (δ, ε, n), such that the intersection of K and T_1 is $\gamma D(T)$-stable modulo L.

(c) γ is also a lower bound for all $\tau(T_t), (D(T)/D(T_t))^{\pm 1}$.

To verify 3.27 we proceed by induction over the sequence

$$\cdots \subset T^{r,j} \subset T^{r,j+1} \subset T^{r,j+2} \subset \cdots$$

which was defined just after 3.22 above. Our induction hypothesis is the same as 3.23(r,j) with the exception that in 3.23(r,j)(b) the intersection of K and T_1 is required to be $\gamma_{r,j} D(T)$-stable modulo L (instead of just $\gamma_{r,j}D(T)$-stable), and the P.L. flow $\varphi_{r,j,t} : T \to T_t, t \in [0,1]$, of 3.23$(r,j)$(a) is required to leave each $e \in L$ invariant.

Our induction step is carried out a bit differently than in the proof of lemma 3.1 so we describe this step in some detail.

Let $e \in J_{j+1,r}$ (see 3.22 for $J_{j+1,r}$); let d be the minimal set in L with $e \subset d$; let W denote the linear polyhedron generated by K and L; let $e_1, e_2, \ldots, e_\beta$ be the members of W with $e_i \subset d$, $e_i \not\subset \partial d$, and $\dim(e_i) + \dim(e) < \dim(d)$ holding for all i. Set $e' = \varphi_{r,j,1}(e)$. By the hypothesis of lemma 3.2 the intersection $e \cap \partial d$ is a simplex; so $e' \cap \partial d$ is also a simplex. Thus e' can be written as $\langle v_0, v_1, \ldots, v_k, v_{k+1}, \ldots, v_p\rangle$ where the $\{v_i\}$ are its vertices and $e' \cap \partial d = \langle v_{k+1}, \ldots, v_p\rangle$. We can choose a P.L. flow $\varphi_{e,t} : e' \to e'_t, t \in [0,1]$, to satisfy all the following properties.

3.28. (a) 3.24(a); $\varphi_{e,t} : e' \to e'_t$, $t \in [0,1]$, does not move any of the vertices $\langle v_{k+1}, v_{k+2}, \ldots, v_p\rangle$.

(b) There is a number $\eta > 0$, which depends only on the $(\gamma_{r,j}, \beta, n, \varepsilon)$ such that for all i the distance from $\varphi_{e,1}(\langle v_0, v_1, \ldots, v_k, \rangle)$ to Q_i is greater than $\eta D(T)$. Here Q_i is the smallest plane in R^n containing both e_i and $\langle v_{k+1}, \ldots, v_p\rangle$.

(c) $e'_t \subset d$ for all $t \in [0,1]$.

Towards verifying 3.28 we note that $\dim(Q_i) + \dim(\langle v_0, v_1, \ldots, v_k)\rangle < \dim(d)$, so by varying the positions of the vertices of $\langle v_0, v_1, \ldots, v_k\rangle$ a little bit (this variation is the flow $\varphi_{e,t} : e' \to e'_t$) we can move $\langle v_0, v_1, \ldots, v_k\rangle$ away from all the Q_i. The remaining details in the verification of 3.28 are left as an exercise.

Using the $\varphi_{e,t} : e' \to e'_t, t \in [0,1]$, of 3.28 we can define $\varphi_{r,j+1,t} : T \to T_t, t \in [0,1]$, by 3.25. Note that the P.L. flow $\varphi_{r,j+1,t} : T \to T_t, t \in [0,1]$, satisfies 3.23$(r, j+1)$, provided the η and the $\varphi_{e,t} : e' \to e'_t$, $t \in [0,1]$, of 3.28 are chosen sufficiently close to 0 and the identity map.

This completes the proof of lemmas 3.2, 3.3.

PROOF OF LEMMA 3.4: In this proof we assume $|K| \subset |L|$. The proof without this assumption is carried out in the same way. Note from the hypothesis $K\#D(L) \leq \gamma$ of 3.4 it follows that there is a positive integer m and, for each $q \in \{0, 1, \ldots, n\}$, subsets $J_{q,i} \subset K^q - K^{q-1}$ $(1 \leq i \leq m)$ which satisfy the following.

3.29 (a) $K^q - K^{q-1} = \cup_{i=1}^m J_{q,i}$; $J_{q,i} \cap J_{q,j} \neq \emptyset$ if and only if $i = j$.

(b) For any $e \in L$ and any q, i we have that e intersects at most one of the sets in $J_{q,i}$.

(c) There is an upper bound for m which depends only on the (γ, n).

Set $K^{q,i} = K^q \cup (\cup_{j=1}^{i} J_{q+1,j})$. The proof of lemma 3.4 proceeds by induction over the sequence $\ldots, K^{q,i}K^{q,i+1}, \ldots$. Here is the induction hypothesis.

<u>3.30(q,i)</u> There is an s-fold derived subdivision $L^{(s)}$ of L which satisfies the following:

(a) There is a subcomplex $C \subset L^{(s)}$ such that $|C| = |K^{q,i}|$. (Recall the assumption $|K| \subset |L|$.) Moreover for each $e \in K^{q,i}$ and each $e' \in L$ we have that $e \cap e'$ is a subcomplex of C.

(b) $L^{(s)}$ is in transverse position to K modulo $K^{q,i}$. Let V denote the linear polyhedron generated by K and $L^{(s)}$.

(c) There is a lower bound for all the $\tau(L^{(s)}), \tau(V)$, $D(L^{(s)})/D(L)$, $D(V)/D(L)$ which depends only on the (δ, q, i, n).

(d) There is an upper bound for the integer s which depends only on the (γ, q, i, n).

Our induction step consists of showing $3.30(q,i) \Rightarrow 3.30(q,i+1)$ (or of showing $3.30(q,i) \Rightarrow 3.30(q+1,1)$ if $i = m$.)

Define a derived subdivision of $L^{(s)}, L^{(s+1)}$, by adding new vertices to $L^{(s)}$ of 3.30(q,i), one new vertex $v(e)$ on the interior of each triangle $e \in L^{(s)}$, in the following manner. If for $e \in L^{(s)}$ there is $e' \in J_{q+1,i+1}$ such that $(e - \partial e) \cap e' \neq \emptyset$ then let $v(e)$ denote the geometric center of the convex cell $e \cap e'$. Otherwise choose $v(e)$ to be the geometric center of e. Next choose a derived subdivision $L^{(s+2)}$ of $L^{(s+1)}$ subject to the following restrictions. If for $e \in L^{(s+1)}$ we denote by $v(e)$ the vertex of $L^{(s+2)}$ which is contained in $e - \partial e$, then $v(e)$ is chosen as the geometric center of e if e does not intersect the interior of any $e' \in J_{q+1,i+1}$. On the other hand if $e \cap (e' - \partial e') \neq \emptyset$ for some $e' \in J_{q+1,i+1}$ then we choose $v(e)$ to be sufficiently close to e' so as to assure that the following is satisfied:

<u>3.31</u> (a) If $f \in L^{(s+2)}$ satisfies $f \cap (e' - \partial e') \neq \emptyset$ for some $e' \in J_{q+1,i+1}$, and $f \cap (e'' - \partial e'') \neq \emptyset$ for some $e'' \in K - K^{q,i}$, then we must have that $e' \subset e''$.

Next we apply Lemma 3.2 first to $K, L^{(s)}$ and $L^{(s+1)}$, and next to $K, L^{(s+1)}$, and $L^{(s+2)}$ to obtain a P.L. flow of $L^{(s+2)}$ which leaves the triangles of $L^{(s)}$ invariant and which moves $L^{(s+2)}$ to a new 2-fold derived subdivision of $L^{(s)}$—which we also denote by $L^{(s+2)}$—satisfying the following.

<u>3.31</u> (b) $L^{(s+2)}$ is in transverse position to K modulo $K^{q,i}$.

It is left as an exercise to show that the $L^{(s+2)}$ of 3.31(b) can be chosen so that 3.31(a) is still satisfied and so that the following holds.

3.31 (c) Let W denote the linear polyhedron generated by $L^{(s+2)}$ and K. Then there is a lower bound for all $\tau(W)$, $\tau(L^{(s+2)})$, $D(W)/D(L)$, $D(L^{(s+2)})/D(L)$ which depends only on (δ, q, i, n).

To complete the induction step $3.30(q,i) \Rightarrow 3.30(q,i+1)$ we still need to subdivide once more. For each $e \in L^{(s+2)}$ add a new vertex $v(e)$ in $e - \partial e$ as follows: $v(e)$ is the geometric center of e if e intersects the interior of no set in $J_{q+1,i+1}$; if $e \cap (e' - \partial e') \neq \emptyset$ for some $e' \in J_{q+1,i+1}$ then set $v(e)$ equal the geometric center of the convex cell $e \cap e'$. These new vertices define $L^{(s+3)}$—a derived subdivision of the $L^{(s+2)}$ in 3.31. Note that $L^{(s+3)}$ already satisfies 3.30(q,i+1)(a)(d). However $L^{(s+3)}$ may not satisfy 3.30(q,i+1)(b)(c). To remedy this we must vary some of the $v(e)$ through paths $v(e,t)$, $t \in [0,1]$, in $e - \partial e$. Note that 3.31 and the construction of $L^{(s+3)}$ (thus far) allow us to apply lemma 3.2 to $K, L^{(s+2)}$, and $L^{(s+3)}$ to get a vertex-wise flow $\varphi_t : L^{(s+3)} \to L_t^{(s+3)}$, $t \in [0,1]$, which leaves each triangle of $L^{(s+2)}$ invariant and satisfies:

3.32 (a) $L_1^{(s+3)}$ is in transverse position to K modulo $K^{q,i}$. Let W denote the linear polyhedron generated by $L_1^{(s+3)}$ and K.

(b) There is a lower bound for all of $\tau(W)$, $\tau(L_1^{(s+3)})$, $D(W)/D(L)$, $D(L_1^{(s+3)})/D(L)$ which depends only on (δ, q, i, n).

Now define another derived subdivision of the $L^{(s+2)}$ of 3.31—this will also be denoted by $L^{(s+3)}$—by adding the following vertices to those of $L^{(s+2)}$: if e doesn't intersect the interiors of any sets in $J_{q+1,i+1}$ then add $\varphi_1(v(e))$ as the new vertex in $e - \partial e$; if $e \cap (e' - \partial e') \neq \emptyset$ for some $e' \in J_{q+1,i+1}$ then add $v(e)$ as the new vertex in $e - \partial e$ ($v(e)$ is the geometric center of $e \cap e'$).

It is left as an exercise to deduce from 3.31, 3.32, 3.30(q,i), and the construction of this final $L^{(s+3)}$, that $L^{(s+3)}$ satisfies 3.30(q,i+1).

Note that 3.4(a)-(c) follow from 3.30(n,1). The verification of 3.4(d) is left as an exercise.

This completes the proof of lemma 3.4.

PROOF OF LEMMA 3.7.: For any continuous map $r : X \to R^n$ (with $X \subset R^n$) there is the following definition.

DEFINITION 3.33. *We write $C^1(r) < \beta$—for some $\beta > 0$—if and only if the following hold.*

(a) For all $x \in X$ we have $|r(x) - x| < \beta$.

(b) For any $x_1, x_2 \in X$ we must have that

$$|r(x_2) - r(x_1) - (x_2 - x_1)| < \beta |x_2 - x_1|.$$

(Note that 3.33 is an attempt to define the notion of $r : X \to R^n$ being close to 1_X in the C^1-topology—even though $r : X \to R^n$ may not be differentiable.)

Note that the hypothesis $|\varphi_{i,t}(x) - x| < \varepsilon D(T_i)$ of 3.7 implies that the following hold.

3.34. There is $\varepsilon_1 > 0$ which satisfies the following.

(a) $C^1(\varphi_{i,1}) < \varepsilon_1 D(T_i)$ holds for all $i \in \{1, 2, \ldots, u\}$.

(b) ε_1 depends only on (δ, ε) of 3.7.

(c) $\lim_{\varepsilon \to 0} \varepsilon_1 = 0$.

In the remainder of this proof we use only the properties 3.34 and 3.6(e) to deduce that there is a homeomorphism $g : R^n \to R^n$ as in 3.7(a)(b).

We first prove lemma 3.7 when hypothesis 3.35 is satisfied. Then we briefly discuss how to prove 3.7 when hypothesis 3.35 is dropped.

Hypothesis 3.35. Suppose that $\varphi_t : R^n \to R^n$ is the identity map for each $t \in [0, 1]$. (Here φ_t comes from 3.6.)

Let W denote the linear polyhedron generated by all the $\{L_i : 1 \leq i \leq u\}$ and K. For each $e \subset W$ we choose a P.L. collaring $c_e : \partial e \times [0, 1] \to e$ for ∂e in e. For each $t \in [0, 1]$ we set ${}_te = e - c_e(\partial e \times [0, t])$. We also choose for each $e \in W$ an embedding $h_e : {}_{1/4}e \times R^{n-\dim(e)} \to R^n$. The $\{c_e, h_e : e \in W\}$ can be chosen to satisfy the following.

3.36. (a) $c_e(q, 0) = q$ for all $q \in \partial e$ and all $e \in W$. $h_e(q, 0) = q$ for all $q \in {}_{1/4}e$ and all $e \in W$. For each $e \in W$ the restriction $h_e|({}_{1/4}e \times {}_{10}B^{n-\dim(e)})$ is a linear map. (For any $t > 0$ we let ${}_tB^{n-\dim(e)}$ denote the ball of radius t centered at the origin in $R^{n-\dim(e)}$.)

(b) For any $e, e' \in W$ we have $\text{Image}(h_e) \cap \text{Image}(h_{e'}) \neq \emptyset$ only if $e \subset e'$ or $e' \subset e$.

(c) For each $e \in W$ let $e_1, e_2, \ldots, e_y$ denote all the minimal sets of $K \cup (\cup_{i=1}^u L_i)$ which satisfy $e \subset e_i$. We require that there is a vector subspace $A_{e,e_i} \subset R^{n-\dim(e)}$ such that $h_e({}_{1/4}e \times A_{e,e_i}) = \text{Image}(h_e) \cap e_i$, for each $i \in \{1, 2, \ldots, y\}$.

(d) For any $s \in \{-1, 0, 1, 2, \ldots, n\}$ set

$$N_{-1} = \emptyset$$

$$N_s = \left[N_{s-1} - \bigcup_{e \in W^s - W^{s-1}} h_e({}_{1/4}e \times {}_2B^{n-\dim(e)}) \right] \bigcup \left[\bigcup_{e \in W^s - W^{s-1}} h_e({}_{1/4}e \times {}_1B^{n-\dim(e)}) \right].$$

For any $e \in W^{s+1} - W^s$ we have $h_e(({}_{1/4}e - {}_{1/2}\, e) \times {}_1B^{n-\dim(e)}) \subset N_s$; for any $e \in W - W^s$ we have $N_s \cap h_e({}_1e \times {}_1B^{n-\dim(e)}) = \emptyset$.

Before defining the map $g : R^n \to R^n$ of 3.7 we need to define some preliminary maps $r_e : h_e({}_{1/3}e \times {}_1B^{n-\dim(e)}) \to R^n$, one such map for each $e \in W$, as follows. Let $e_1, e_2, \ldots, e_y$ be as in 3.36(c), and for each $q \in {}_1B^{n-\dim(e)}$ denote by $A_{e,e_i}(q)$ the image in $R^{n-\dim(e)}$ of A_{e,e_i} under translation by q.

3.37. For any $(p, q) \in {}_{1/3}e \times {}_1B^{n-\dim(e)}$ we set

$$r_e(h_e(p,q)) = h_e(p \times R^{n-\dim(e)}) \cap \left(\cap_{i=1}^{y} s_i \circ h_e({}_{1/4}e \times A_{e,e_i}(q))\right).$$

Here $s_i = \varphi_{i',1}$ if $e_i \in L_{i'}$, or $s_i = 1$ if $e_i \in K$. We note that it follows from 3.34 that if ε_1 is sufficiently small in 3.34 then the $r_e : h_e({}_{1/3}e \times {}_1B^{n-\dim(e)}) \to R^n$ of 3.37 are well defined embeddings.

We leave as an exercise the verification of the following properties of the embeddings $\{r_e : e \in W\}$.

3.38. (a) There is $\varepsilon_2 > 0$ such that for any $e \in W$ we must have $C^1(r_e) < \varepsilon_2 D$, where $D = \text{minimum}\{D(T_i) : 1 \le i \le u\}$.

(b) $\lim_{\varepsilon_1 \to 0} \varepsilon_2 = 0$.

(c) For any $e, e' \in W$ with $e' \subset e$ and for any $b \in {}_1B^{n-\dim(e)}$, we must have that $r_e^{-1} \circ r_{e'}(h_e({}_{1/4}e \times b)) \subset h_e({}_{1/4}e \times b)$ and $r_{e'}^{-1} \circ r_e(h_e({}_{1/4}e \times b)) \subset h_e({}_{1/4}e \times b)$. (Note that $r_e^{-1} \circ r_{e'}, r_{e'}^{-1} \circ r_e$ may not be defined on all of $h_e({}_{1/4}e \times b)$.)

In the remainder of this proof the homeomorphism $g : R^n \to R^n$ of 3.7 will be constructed from the $\{r_e : e \in W\}$ by an induction argument carried out over the skeleta of W.

Here is our induction hypothesis.

3.40(s). There is an embedding $g_s : N_s \to R^n$ which satisfies the following properties.

(a) There is $\varepsilon_{2+s} > 0$ such that we have $C^1(g_s) < \varepsilon_{2+s} D$, where D comes from 3.38(a).

(b) $\lim_{\varepsilon_2 \to 0} \varepsilon_{2+s} = 0$.

(c) For any $e \in W - W^s$ and any $b \in {}_1B^{n-\dim(e)}$ we must have $r_e^{-1} \circ g_s(h_e({}_{1/4}e \times b)) \subset h_e({}_{1/4}e \times b)$ and $g_s^{-1} \circ r_e(h_e({}_{1/4}e \times b)) \subset h_e({}_{1/4}e \times b)$. For any $e \in W$, and any $b \in {}_1B^{n-\dim(e)}$ sufficiently close to the origin, we must have $r_e^{-1} \circ g_s(h_e({}_{1/4}e \times b)) \subset h_e({}_{1/4}e \times b)$ and $g_s^{-1} \circ r_e(h_e({}_{1/4}e \times b)) \subset h_e({}_{1/4}e \times b)$. (Note that $g_s^{-1} \circ r_e, r_e^{-1} \circ g_s$ may not be defined on all of $h_e({}_{1/4}e \times b)$.)

Our induction step consists of showing 3.40(s) $\Rightarrow$ 3.40($s+1$). We would like to union g_s with all the $\{r_e : e \in W^{s+1} - W^s\}$ to get g_{s+1}. Unfortunately this does not yield a well defined map because the $\{r_e : e \in W^{s+1} - W^s\}$ and g_s are not necessarily equal on their overlapping domains. They are however all very close to the identity map if ε of 3.7 is chosen sufficiently small (see 3.34(c), 3.38(a)(b), 3.40(s)(a)(b)). Because of this closeness, the formulae of 3.41(a)(b) yield for each $e \in W^{s+1} - W^s$ a new embedding $r'_e : h_e({}_{1/3}e \times {}_1B^{n-\dim(e)}) \to R^n$ which satisfies 3.41(c)-(e).

<u>3.41.</u> (a) Define $f_e : h_e(({}_{2/5}e - {}_{3/7}\, e) \times {}_1B^{n-\dim(e)}) \to R^n$ as follows:

$$f_e(h_e(c_e(x,t),b)) = t_1 h_e(c_e(x,t),b) + t_2 g_s^{-1} \circ r_e(h_e(c_e(x,t),b)),$$

where $(x,t) \in \partial e \times [2/5, 3/7]$, $b \in {}_1B^{n-\dim(e)}$, $t_1 = (2/5 - 3/7)^{-1}(t - 3/7)$, and $t_2 = (3/7 - 2/5)^{-1}(t - 2/5)$.

(b) Set r'_e equal to g_s on $h_e(({}_{1/3}e - {}_{2/5}\, e) \times {}_1B^{n-\dim(e)})$; Set $r'_{e'}$ equal to $g_s \circ f_e$ on $h_e(({}_{2/5}e - {}_{3/7}\, e) \times {}_1B^{n-\dim(e)})$; Set r'_e equal to r_e on $h_e({}_{3/7}e \times {}_1B^{n-\dim(e)})$.

(c) There is $\varepsilon_{3+s} > 0$ such that $C^1(r_e{}') < \varepsilon_{3+s} D$.

(d) $\lim_{\varepsilon_2 \to 0} \varepsilon_{3+s} = 0$; $\varepsilon_{3+s} \geq \varepsilon_{2+s}$.

(e) For any $e' \in W - W_s$ and any $b \in {}_1B^{n-\dim(e)}$ we must have that $r_{e'}^{-1} \circ r'_e(h_{e'}({}_{1/4}e' \times b)) \subset h_{e'}({}_{1/4}e' \times b)$ and ${r'_e}^{-1} \circ r_{e'}(h_{e'}({}_{1/4}e' \times b)) \subset h_{e'}({}_{1/4}e' \times b)$. (Note that $r_{e'}^{-1} \circ r'_e, {r'_e}^{-1} \circ r_{e'}$ may not be defined on all of $h_{e'}({}_{1/4}e' \times b)$.)

We can now define $g_{s+1} : N_{s+1} \to R^n$ as follows.

<u>3.42.</u> Set g_{s+1} equal to g_s on $N_s - (\cup_{e \in W^{s+1} - W^s} \text{domain}(r'_e))$; for each $e \in W^{s+1} - W^s$ set g_{s+1} equal to r'_e on domain(r'_e). It now follows from 3.40(s), 3.41 that the $g_{s+1} : N_{s+1} \to R^n$ defined in 3.42 satisfies 3.40($s+1$) for the $\varepsilon_{2+(s+1)}$ given in 3.41(c).

This completes the induction step in our proof of 3.40(n). Set $g : R^n \to R^n$ equal to g_n on N_n. Note that g_n equals 1 on $N_n - \cup_{i=1}^u |L_i|$, where the $\{L_i\}$ come from 3.6 (see 3.6(e) and review the construction of g_n). Thus by setting g equal to 1 on $R^n - N_n$ we have a homeomorphism $g : R^n \to R^n$ which satisfies 3.7(a). That $g : R^n \to R^n$ satisfies 3.7(b) follows from 3.36, 3.37, 3.40(n)(c), and the definition of $g : R^n \to R^n$.

To complete the proof of 3.7 it remains to show that ε of 3.7 depends only on (δ, n, u). (Thus far we know only that for sufficiently small ε lemma 3.7 will be true.) The verification of this dependence is left as an exercise. We note that it will be necessary to assume that the $\{h_e, c_e : e \in W\}$ of 3.36 satisfy further "metric properties."

This completes the proof of lemma 3.7 for the special case when the hypothesis 3.35 is satisfied.

We now briefly discuss how to prove lemma 3.7 when hypothesis 3.35 is not satisfied. To begin with we must replace the map h_e of 3.36 by

$$\overline{h}_e = \varphi_1 \circ h_e,$$

where $\varphi_1 : R^n \to R^n$ is as in 3.6. Now by replacing h_e in 3.37 by $\overline{h}_e$ the folmula of 3.37 yields an embedding

$$\overline{r}_e : \overline{h}_e({}_{1/3}e \times {}_1B^{n-\dim(e)}) \to R^n,$$

provided ε_1 of 3.34 is sufficiently small. The embeddings $\{\overline{r}_e\}$ satisfy 3.38(c) (when in 3.38(c) we replace the $r_e, r_{e'}, h_e$ by $\overline{r}_e, \overline{r}_{e'}, \overline{h}_e$.) The $\{\overline{r}_e\}$ do not in general satisfy 3.38(a)(b), but do satisfy a weakened form of 3.38(a)(b). Before stating this weakened form of 3.38(a)(b) we need some more notation.

Let $f : X \to R^n$ be a continuous map with $X \subset R^n$. We shall write

$$\overline{C}^1(f) < \varepsilon$$

if the following hold true.

3.43. (a) $|f(x) - x| < \varepsilon$ for all $x \in X$.

(b) For any $e \in W$ let $e_1, e_2, \ldots, e_y$ be as in 3.36(c). We suppose that e_1 is the only one of the $e_1, e_2, \ldots, e_y$ contained in K ($e_1 = \emptyset$ is allowed). Let $A_{e,e_1}(q)$ be as in 3.37 for $q \in R^{n-\dim(e)}$. Then for any $q \in {}_1B^{n-\dim(e)}$ and any $x, y \in X \cap \overline{h}_e({}_{1/4+\varepsilon}e \times ({}_1B^{n-\dim(e)} \cap A_{e,e_1}(q)))$ we must have

$$|f(x) - f(y) - (x - y)| < \varepsilon |x - y|,$$

and

$$f(x), f(y) \in \overline{h}_e({}_{1/4}e \times ({}_1B^{n-\dim(e)} \cap A_{e,e_1}(q))).$$

(c) If in (b) we have that $e_1 = \emptyset$ then we require that the restriction $f|X_e$—where $X_e = X \cap \overline{h}_e({}_{1/4}e \times {}_1B^{n-\dim(e)})$—satisfy $C^1(f|X_e) < \varepsilon$.

We can now state the weakened version of 3.38(a)(b) which the $\{\overline{r}_e\}$ satisfy.

3.38. ($\overline{a}$) There is $\varepsilon_2 > 0$ such that for any $e \in W$ we must have $\overline{C}_1(\overline{r}_e) < \varepsilon_2 D$, where $D =$ minimum$\{D(T_i) : 1 \leq i \leq u\}$.

($\overline{b}$) $\lim_{\varepsilon_1 \to 0} \varepsilon_2 = 0$.

We use the $\{\overline{r}_e\}$ to construct the $g_s : N_s \longrightarrow R^n$ of 3.40(s) (instead of using the $\{r_e\}$). The induction hypothesis 3.40(s) must be slightly altered. In 3.40(s)(c) the r_e, h_e must be replaced by the $\{\overline{r}_e\}, \{\overline{h}_e\}$; and 3.40(s)(a)(b) must be replaced by the following weaker conditions.

3.40(s) ($\overline{a}$) There is $\varepsilon_{2+s} > 0$ such that $\overline{C}^1(g_s) < \varepsilon_{2+s} D$.
($\overline{b}$) $\lim_{\varepsilon_2 \to 0} \varepsilon_{2+s} = 0$.

If in 3.41(a)(b) we replace r_e, h_e by $\overline{r}_e, \overline{h}_e$ we get an embedding

$$\overline{r}'_e : \overline{h}_e({}_{1/3}e \times_1 B^{n-\dim(e)}) \longrightarrow R^n.$$

(Note that 3.38($\overline{a}$)($\overline{b}$), 3.40(s)($\overline{a}$)($\overline{b}$), 3.38(c), and 3.40(s)(c) assure that $\overline{r}'_e$ is well defined by the formulae of 3.41(a)(b) provided ε_2 is sufficiently small.) The $\overline{r}'_e$ satisfy 3.41(e) if in 3.41(e) we replace $h_{e'}$, r'_e, $r_{e'}$ by $\overline{h}_{e'}$, $\overline{r}'_e$, $\overline{r}_{e'}$. The $\overline{r}'_e$ satisfy the following weakened form of 3.41(c)(d).

3.41. ($\overline{c}$) There is $\varepsilon_{3+s} > 0$ such that $\overline{C}^1(r'_e) < \varepsilon_{3+s} D$.
($\overline{d}$) $\lim_{\varepsilon_2 \to 0} \varepsilon_{3+s} = 0$; $\varepsilon_{3+s} > \varepsilon_{2+s}$.

Finally we carry out the induction step 3.40(s) $\Rightarrow$ 3.40(s+1) by defining $g_{s+1} : N_{s+1} \longrightarrow R^n$ as in 3.42 (where r'_e is replaced by $\overline{r}'_e$ in 3.42).
This completes the proof of lemma 3.7.

§4. SOME SMOOTH CONSTRUCTIONS

In this section piecewise smooth versions of the constructions of section 2 are discussed. In particular piecewise smooth versions of propositions 2.10 and 2.14 are stated (see 4.8, 4.9). The techniques of piecewise smooth triangulation theory (as exposited in [18]) assure that the proofs given in section 3 for propositions 2.10 and 2.14 carry over directly to give proofs for the corresponding propositions of this section.

We begin by rephrasing all the definitions of section 2 in the smooth category.

A continuous map $f : X \to R^n$ from a subset $X \subset R^n$ is called a **smooth immersion** if there is an open neighborhood Y for X in R^n and an extension of $f : X \to R^n$ to $F : Y \to R^n$ such that F is a smooth immersion in the usual sense.

DEFINITION 4.1. *A* **smooth polyhedron** *K in R^n consists of a collection of subsets $\{e \in K\}$ of R^n for which there exists a homeomorphism $h : R^n \to R^n$ and a linear polyhedron L in R^n satisfying the following:*

(a) $e \in K$ only if there is $e' \in L$ such that $h(e') = e$.

(b) For each $e' \in L$ the mapping $h_{|e'}$ is a smooth immersion.

(c) $|K| = h(|L|)$. (Here $|K|$ is the union of all $e \in K$.)

The pair (h, L) of 4.1 is called a **linear parametrization** for the smooth polyhedron K of 4.1. Note that 4.1 does not uniquely specify a linear parametrization for K, but rather just states that one exists. For e, e', h as in 4.1 set $\partial e = h(\partial e')$.

A smooth polyhedron K is called **full** if it has a linear parametrization $(h, L))$ such that L is full. We call K a **smooth triangulation** if it has a linear parametrization (h, L) where L is a linear triangulation in R^n. A **subdivision** of K consists of another smooth polyhedron M such that $|M| = |K|$ and each $e \in K$ is a union of sets in M. A **subcomplex** of K is a subset $K' \subset K$ such that K' is a smooth polyhedron. We denote by K^i the subcomplex consisting of all $e \in K$ such that $\dim(e) \leq i$.

DEFINITION 4.2. *A subdivision M of a smooth triangulation K in R^n is called an* **r-fold derived subdivision** *of K, and denoted by $K^{(r)}$, if there is a linear parametrization (h, L) for K, an r-fold derived subdivision $L^{(r)}$ of L, and a linear parametrization $(h', L^{(r)})$ for M such that $h'(|f|) = h(f)$ for all $f \in L$.*

Let $\{K_i : 1 \leq i \leq u\}$ denote a finite collection of full smooth polyhedra in R^n. We can define the structure **generated** by the $\{K_i : 1 \leq i \leq u\}$ by using the requirements of 2.5(a)(b), and denote this generated structure by W. Note that W is a finite collection of compact subsets of R^n, but W need not be a smooth polyhedron. However in the following definitions we always require that W be a smooth polyhedron.

DEFINITION 4.3. *The full smooth polyhedra $\{K_i : 1 \leq i \leq u\}$ are in* **transverse position** *to one another in R^n if the following hold:*

(a) For every selection $e_i \in K_i$, $1 \leq i \leq u$, the smooth manifolds $\{(e_i - \partial e_i) : 1 \leq i \leq u\}$ are in transverse position to one another in R^n (transverse in the smooth category).

(b) W (the structure generated by the $\{K_i : 1 \leq i \leq u\}$) is a smooth polyhedron.

DEFINITION 4.4. *Two smooth polyhedra K_1, K_2 in R^n are in* **transverse position** *to one another in R^n* **modulo** *a third smooth polyhedron K_3 if the following hold:*

(a) 2.7(a)(b)(c) are true in the smooth category (i.e. " transverse" must mean transverse in the smooth category).

(b) The structure generated by K_1 and K_2 is a smooth polyhedron.

In the next definition B denotes a finite regular cell complex, A denotes a subcomplex of B, and K_A denotes a collection of full smooth polyhedra in R^n—one smooth polyhedron K_e for each $e \in A$. For any subcomplex $C \subset A$ let $K(C)$ denote the structure generated by all the $\{K_e : e \in C\}$.

DEFINITION 4.5. *The collection K_A of full smooth polyhedra are in B-transverse position to one another in R^n if the following are true:*

(a) For any $e \in B$ and any subcomplex C of $e \cap A$ the structure $K(C)$ is a smooth polyhedron.

(b) For any $e \in B$ and any two subcomplexes C_1, C_2 of $e \cap A$ we have that $K(C_1)$ and $K(C_2)$ are in transverse position to one another modulo $K(C_1 \cap C_2)$.

METRIC NOTATION 4.6. For any smooth polyhedron K the definitions of 2.9 can be used to define the $D(K), \overline{D}(K)$ and the thickness $\tau(K)$. For any A, B, K_A as in 4.5 $D(K_A)$ and thickness $\tau(K_A)$ can also be defined as in 2.9.

DEFINITION 4.7. *A smooth polyhedron K is called an ε-linear polyhedron (for some $\varepsilon > o$) if there is a linear parametrization (h, L) for K satisfying:*

(a) $|p - h(p)| \leq \varepsilon D(K)$ for all $p \in |L|$.

(b) For any $e \in L$ and any non-zero vector $v \in T(e)$ we have that $|v - d(h_{|e})(v)| \leq \varepsilon |v|$. Here $d(h_{|e})$ is the differential of the smooth immersion $h_{|e}$, and $v, d(h_{|e})(v)$ have been translated to the origin.

We can now state the smooth version of proposition 2.10.

PROPOSITION 4.8. *Let r denote a given positive integer, let δ be a given positive number, and let B be a given finite regular cell complex. Then given any $t > 0$ there is $s > 0$, which depends only on the numbers $(r, t, n, \delta, N(B))$, such that the following is true. Suppose A, B, K_A are as in 4.5, with $\tau(K_A) > \delta$, and suppose that for each $e \in B$ and any subcomplex C of $e \cap A$ we have that $K(C)$ is an s-linear polyhedron. Then there is an extension of K_A to a collection K_B of smooth polyhedra which satisfy the smooth versions of the conclusions of 2.10. Moreover for each $e \in B$ and any subcomplex C of e we have that $K(C)$ is a t-linear polyhedron.*

The notion of a finite collection $\{K_i : i \in I\}$ of smooth polyhedra being in **α-transverse position** to one another (for a positive number α) is defined exatly as in 2.12. For any number $\varepsilon > 0$ and any smooth polyhedron K in R^n the integer $K\#\varepsilon$ is defined exactly as in 2.11.

We can now state the smooth version of proposition 2.14.

PROPOSITION 4.9. *Let $b \geq n$ denote a given positive integer and let δ, γ denote given positive numbers. Then for any $t > 0$ there is $s > 0$, which depends only on the numbers $(t, b, n, \delta, \gamma)$, such that the following is true. Let L, K $\{K_i : 1 \leq i \leq b\}$, $\{K_{i,j} : 1 \leq i \leq n, j \in J_i\}$ denote smooth polyhedra in R^n which satisfy smooth versions of the hypothesis of 2.14 (see 2.13). In addition we suppose that any smooth polyhedron generated by any subcollection of the collection $K, L, \{K_{i,j} : 1 \leq i \leq n, j \in J_i\}$, and any smooth polyhedron generated by any subcollection of the $\{K_{n,j} : j \in J_n\} \cup \{K_k\}$ (here we have a different collection for each $k \in \{n+1, n+2, \ldots, b\}$), are all s-linear polyhedra. There is a positive number β which depends only on (n, γ). For any β-fold derived subdivision $L^{(\beta)}$ of L, such that $L^{(\beta)}$ is also an s-linear polyhedron and satisfies $\delta < \tau(L^{(\beta)}), \delta < D(L^{(\beta)})/D(L)$, there are homeomorphisms $f_i : R^n \to R^n, i = 1, 2, \ldots, n+1$, which satisfy smooth versions of the conclusions of 2.14, for the given δ, γ. Moreover for each $k \in \{n+1, \ldots, b\}$ the smooth polyhedron generated by any subcollection of the $\{K'_{i,j} : 1 \leq i \leq n, j \in J_i\}$ and K_k is a t-linear polyhedron.*

PROOF OF PROPOSITIONS 4.8, 4.9: The techniques of piecewise smooth triangulation theory—as exposited in [18]—allow the extension of lemmas 3.1, 3.2, 3.3, 3.4, 3.7 to ε-linear versions. Then 4.8, 4.9 are proven exactly as were propositions 2.10, 2.14 (using the ε-linear versions of 3.1–3.4, 3.7 instead of their original linear versions).

We note that lemmas 3.1, 3.2, 3.3, 3.7 deal with P.L. flows of linear polyhedra, a concept which has not yet been discussed for smooth polyhedra.

DEFINITION 4.10. *Let K denote a smooth triangulation in R^n. A P.S. flow of K consists of a linear parametrization (h, L) for K and an isotopy $\varphi_t : R^n \to R^n$, $t \in [0,1]$, of $h : R^n \to R^n$ satisfying the following:*

(a) For each $e \in L$ the mapping $g : e \times [0,1] \to R^n \times R$ is a one-one smooth immersion, where $g(x,t) = (\varphi_t(x), t)$.

(b) For each $t \in [0,1]$ define K_t by: $e \in K_t$ if and only if there is $e' \in L$ with $\varphi_t(e') = e$. Then K_t is a smooth triangulation in R^n with linear parametrization (φ_t, L).

In the ε-linear versions of lemmas 3.1, 3.2, 3.3, 3.7 we begin with ε-linear triangulations and polyhedra which have a specified lower bound $\gamma > 0$ for their thicknesses and the ratios of their diameters. Any P. S. flow $\varphi_t : L \to K_t, t \in [0,1]$, is required to have each level K_t an ε'-linear triangulation with respect to the linear parametrization $(\varphi_t, L))$, where ε' depends only on (ε, n, γ) and $\varepsilon' \to 0$ as $\varepsilon \to 0$.

This completes the proof of 4.8, 4.9.

REMARK 4.11. If a statement is true for all ε-linear polyhedra if ε is sufficiently small then we say that the same statement is true for **almost linear polyhedra.** Likewise if there exist ε-linear polyhedra which satisfy a property for arbitrarily small $\varepsilon > 0$ then we way that there exist almost linear polyhedra which satisfy the property. For example if in 4.8 we replace "s-linear polyhedron" and "t-linear polyhedron" by "almost linear polyhedron" (and suppress all mention of s and t) then we get a restatement of proposition 4.8. The same changes lead to a restatement of proposition 4.9.

§5. THE FOLIATION HYPOTHESIS

Let $F : M \to M$, $\Lambda \subset M$ be as in 1.5. In this section we prove two propositions which concern the nature of $F : M \to M$ near the hyperbolic set $\Lambda \subset M$. We also state a hypothesis which is now placed on $F : M \to M$ and which will be assumed to hold in sections 6 through 13. In section 14 we discuss how to obtain the Markov cell results without use of the hypothesis. Our hypothesis takes the form of a condition placed on $F : M \to M$ near the hyperbolic set $\Lambda \subset M$: a neighborhood V for Λ in M is foliated by a pair of transversal F-invariant foliations which are tangent to ξ_s and ξ_u at Λ.

PROPOSITION 5.1. *There is a neighborhood V for Λ in M and an extension of the splitting $T(M)|_\Lambda = \xi_s \oplus \xi_u$ to a continuous splitting $T(M)|_V = \overline{\xi}_s \oplus \overline{\xi}_u$ satisfying:*

(a) If x, $F(x) \in V$ then $dF(\overline{\xi}_{s|x}) = \overline{\xi}_{s|F(x)}$ and $dF(\overline{\xi}_{u|x}) = \overline{\xi}_{u|F(x)}$.

(b) There is $a \in (0,1), \lambda > 1$ so that for any positive integer q and any $x \in M$ such that $F^i(x) \in V$ for all $i \in \{0, 1, \ldots, q\}$ we have:

$$|dF^q(v)| \geq a\lambda^q|v|, \quad \text{for all } v \text{ in } \overline{\xi}_{u|x}$$

$$|dF^q(v)| \leq a^{-1}\lambda^{-q}|v|, \quad \text{for all } v \text{ in } \overline{\xi}_{s|x}.$$

The following hypothesis is assumed to hold through Section 13.

HYPOTHESIS 5.2: Let $V, \overline{\xi}_s, \overline{\xi}_u$ be as in 5.1. There are smooth foliations $\mathcal{F}_s, \mathcal{F}_u$ of $\mathrm{Int}(V)$ which satisfy:

(a) $\mathcal{F}_s, \mathcal{F}_u$ are tangent to $\overline{\xi}_s, \overline{\xi}_u$ respectively.

(b) Let L_s, L_u denote leaves of $\mathcal{F}_s, \mathcal{F}_u$. Then $F(L_s) \cap \mathrm{Int}(V)$ and $F(L_u) \cap \mathrm{Int}(V)$ are open subsets of leaves of $\mathcal{F}_s$ and $\mathcal{F}_u$ respectively.

PROPOSITION 5.3. *For any positive integer q there are arbitrarily small compact neighborhoods V for Λ in M satisfying*

(a) $F^q(F^q(V) \cap \partial V) \subset M - V$,

(b) $F^{-q}(F^{-q}(V) \cap \partial V) \subset M - V$,

where ∂V denotes the topological boundary for V in M.

We first prove proposition 5.3 and then proposition 5.1. The following lemmas will be needed in the proof of 5.3.

LEMMA 5.4. *For any sufficiently small compact neighborhood V for Λ in M there is a positive integer q such that $F^q(V_s) \subset \mathrm{Int}(V)$. (Here $V_s = \cap_{i \leq 0} F^i(V)$, $V_u = \cap_{i \geq 0} F^i(V)$.)*

PROOF OF LEMMA 5.4: Assume that V is small enough so that $V \subset U$ where U comes from 1.1. Now if $F^q(V_s) \cap \partial V \neq \emptyset$ for all $q \geq 0$ there is

$x \in \cap_{q=0}^{\infty}(F^q(V_s) \cap \partial V)$. Note that $x \in V_s$ and $x \in V_u$. But $V_s \cap V_u = \Lambda$, so $x \in \Lambda$ and $x \in \partial V$ which is a contradiction.

This completes the proof of lemma 5.4.

LEMMA 5.5. *There are arbitrarily small compact neighborhoods V for Λ in M such that $F(V_s) \subset Int(V)$.*

PROOF OF LEMMA 5.5: Start with a compact neighborhood U for Λ in M as in 5.4. Set $U_{s,i} = F^i(U_s) \cap \partial U$. We have

$$U_s \supset U_{s,1} \supset U_{s,2} \supset \cdots \supset U_{s,q} = \emptyset$$

for some positive integer q by 5.4. Choose a very small neighborhood U_q for $F^{-q+1}(U_{s,q-1})$ in U. Set $U' = \overline{U - U_q}$. Then we have

$$U'_s \supset U'_{s,1} \supset U'_{s,2} \supset \cdots \supset U'_{s,q} = \emptyset.$$

Proceeding by induction as in the preceding paragraph we eventually get a compact neighborhood V for Λ in M such that $V_{s,1} = \emptyset$. Since $V_{s,1} = F(V_s) \cap \partial V$ we have $F(V_s) \subset \text{Int}(V)$ as desired.

This completes the proof of lemma 5.5.

PROOF OF PROPOSITION 5.3: It will suffice to prove 5.3 for $q = 1$ in 5.3(a)(b). Start with a neighborhood U as in 5.5. Then by 5.5 we have:

5.6. $F(U) \cap U_s \cap \partial U = \emptyset$.

We claim there is a positive integer p such that the following holds:

5.7. For any $x \in F(U) \cap \partial U$ we have $F^i(x) \in M - U$ for some $i \in \{1, \ldots, p\}$.

To verify 5.7 note that if 5.7 were not true then there would be a convergent sequence $\{x_i\}$ in U with $x_i \in F(U) \cap \partial U$ and $F^j(x_i) \in U$ for all $j \in \{0, 1, \ldots, i\}$. Then $\lim_{i\to\infty} x_i \in F(U) \cap \partial U$ and $\lim_{i\to\infty} x_i \in U_s$, which contradicts 5.6.

Now set $U_p = \{x \in F(U) \cap \partial U : F^i(x) \in U \text{ for all } 0 \leq i \leq p-1\}$. Let N_p denote a very small neighborhood for $F^{p-1}(U_p)$ in U. Set $U' = \overline{U - N_p}$. Note that

$$F(U') \cap \partial U' \subset (F(U) \cap \partial U) \cup N_p,$$
$$F(N_p) \subset M - U \subset M - U'.$$

It follows that

5.8. $U'_p = \emptyset$, where U'_p equals $\{x \in F(U') \cap \partial U' : F^i(x) \in U' \text{ for all } 0 \leq i \leq p-1\}$.

The argument of the preceding paragraph provides the induction step (within an induction) argument which will yield a compact neighborhood V in U satisfying $V_2 = \emptyset$, where $V_j = \{x \in F(V) \cap \partial V : F^i(x) \in V$ for all $0 \leq i \leq j-1\}$. Note that $V_2 = \emptyset \Leftrightarrow$ 5.3(a) is satisfied for $q = 1$.

This completes the proof of 5.3(a). Note that 5.3(a) implies 5.3(b).

This completes the proof of proposition 5.3.

PROOF OF PROPOSITION 5.1: Choose a continuous extension of $T(M)|_\Lambda = \xi_s \oplus \xi_u$ to $T(M)|_N = \tau_s \oplus \tau_u$, where N is a small neighborhood of Λ in M (this just uses the Tietze extension theorem). In the rest of this proof V will denote a very small compact neighborhood for Λ in N. By 5.3 we can choose V to satisfy

5.9. $F(F(V) \cap \partial V) \cap V = \emptyset$; $F^{-1}(F^{-1}(V) \cap \partial V) \cap V = \emptyset$.

Note if V is sufficiently small then the pairs of subbundles $(\tau_{s|X}, \tau_{u|X})$ and $(dF(\tau_s)_{|X}, dF(\tau_u)_{|X})$ will be very close to one another—where $X = V \cap F(V)$. Thus for any two compact disjoint subsets $A \subset V - \Lambda$, $B \subset X - \Lambda$ the Tietze extension theorem can be used once again to extend the splitting $T(M)_{|\Lambda \cup A \cup B} = (\xi_s \oplus \xi_u) \cup (\tau_{s|A} \oplus \tau_{u|A}) \cup (dF(\tau_s)_{|B} \oplus dF(\tau_u)_{|B})$ to a continuous splitting $T(M)_{|V} = \xi'_s \oplus \xi'_u$ which is very close to the splitting $T(M)_{|V} = \tau_{s|V} \oplus \tau_{u|V}$. We have a particular A and B in mind: set $A = F^{-1}(F(\partial V) \cap V)$ and set $B = F(A)$. Note it follows from 5.9 that $A \cap B = \emptyset$. Rephrasing for this particular A and B what we have just shown:

5.10. If V is chosen sufficiently small there will be an extension of the splitting $T(M)_{|\Lambda} = \xi_s \oplus \xi_u$ to a continuous splitting $T(M)_{|V} = \xi'_s \oplus \xi'_u$ which satisfies the following:

(a) The pairs of subbundles (ξ'_s, ξ'_u) and $(\tau_{s|V}, \tau_{u|V})$ are arbitrarily close to one another.

(b) $dF(\xi'_s)_{|B} = \xi'_{s|B}$ and $dF(\xi'_u)_{|B} = \xi'_{u|B}$, where $B = F(\partial V) \cap V$.

Now we can define $\overline{\xi}_u$ of 5.1. Define a sequence $\xi'_{u,j}$ of subbundles of $T(M)_{|V}$ as follows:

5.11. (a) $\xi'_{u,1} = \xi'_u$.

(b) Set $V_1 = \overline{V - F(V)}$, $V_j = V - F(V - V_{j-1})$ for all $j > 1$. Set $\xi'_{u,j}$ equal to $\xi'_{u,j-1}$ over V_{j-1}, and set $\xi'_{u,j}$ equal to $dF(\xi'_{u,j-1})$ over $\overline{V - V_{j-1}}$.

Note it follows from 5.10(b) that the $\xi'_{u,j}$ are well defined by 5.11(b). Set

5.12. $\overline{\xi}_u = \lim_{j \to \infty} \xi'_{u,j}$.

It is left as an exercise to deduce from 5.10(a), 5.11 that for V sufficiently small the limit of 5.12 exists and is a continuous subbundle of $T(M)_{|V}$ which extends ξ_u and which satisfies 5.1(a)(b).

We can define $\overline{\xi}_s$ of 5.1 by applying the argument of the preceding paragraphs to (F^{-1}, τ_s). (Note that the u-direction of F^{-1} is the s-direction of F.)

This completes the proof of proposition 5.1.

§6. SMOOTH TRIANGULATION NEAR Λ

Let $\mathcal{F}_s, \mathcal{F}_u$ be as in 5.2. For a fixed but arbitrary positive integer q we choose a finite set $\{g_i : i \in I\}$ of smooth one-one immersions $g_i : R^n \times R^m \to V$ which satisfy:

6.1. (a) For any $a \in R^n, b \in R^m, i \in I$, we have that $g_i(a \times R^m)$, $g_i(R^n \times b)$ are open subsets of leaves in $\mathcal{F}_s, \mathcal{F}_u$ respectively.

(b) Let B^n, B^m denote the unit balls in R^n, R^m. Set $U = \cup_{i \in I} g_i(B^n \times B^m)$. Then U is a neighborhood of Λ in V which may be assumed to be arbitrarily small.

(c) We have that $F^q(F^q(U) \cap \partial(U)) \subset M - U$ and $F^{-q}(F^{-q}(U) \cap \partial(U)) \subset M - U$ (see 5.3).

(d) For each $i, j \in I$ if $g_i(B^n \times B^m) \cap g_j(B^n \times B^m) \neq \emptyset$ then $g_j(B^n \times B^m) \subset g_i({}_{10}B^n \times_{10} B^m)$. (Here for any $t > 0$ we let ${}_tB^k$ denote the ball of radius t centered at the origin in R^k.)

(e) There is an integer $\eta > 1$ which is independent of the integer q. For each $i \in I$ the number of indices $j \in I$ satisfying Image$(g_j) \cap$Image$(g_i) \neq \emptyset$ is less than η. Moreover for any $v \in T({}_{10}B^n \times_{10} B^m)$ we have

$$\eta^{-1}|v| < |dg_i(v)| < \eta|v|,$$

where dg_i is the derivative of the map $g_i : R^n \times R^m \to M$.

The set of smooth charts in 6.1 is referred to throughout the rest of this paper. Any smooth one-one immersion $g : R^n \times R^m \to V$ satisfying 6.1(a) is called a **rectangular chart** for $\mathcal{F}_s, \mathcal{F}_u$. A subset $e \subset V$ is called a **rectangular cell** if there is a rectangular chart $g : R^n \times R^m \to V$ and cells $e_u \subset R^n, e_s \subset R^m$ such that $g(e_u \times e_s) = e$; in this case we set $\partial_u = g(\partial e_u \times e_s)$ and set $\partial_s e = g(e_u \times \partial e_s)$. A regular cell structure C for a region $|C|$ in V is called a **rectangular cell structure** if each cell of C is a rectangular cell. Note that for each $e \in C$ we must have that $\partial_u e$ and $\partial_s e$ are both subcomplexes of C.

LEMMA 6.2. *There is a finite rectangular cell structure C in U—where U is as in 6.1(b)—which satisfies the following:*

(a) Given any small $\delta > 0$ it may be assumed that diameter$(e) \leq \delta$ holds for all $e \in C$.

(b) For any given compact subset $D \subset Int(U)$ we may assume that $|C| \supset D$.

(c) $N(C) \leq b$, where b is a number which depends only on $n + m$ and η in 6.1(e). (See 2.9 for the definition of $N(C)$.)

PROOF OF LEMMA 6.2: For each $i \in I$ choose compact smooth submanifolds $M_{u,i} \subset B^n - \partial B^n$, $M_{s,i} \subset B^m - \partial B^m$ of codimension zero so that the following are satisfied:

6.3. (a) $D \subset \cup_{i \in I} g_i(M_{u,i} \times M_{s,i})$.

(b) For each $i, j \in I$ set $M_{u,i,j} = \rho_i \circ g_i^{-1} \circ g_j(M_{u,j} \times B^m)$ if $g_i(B^n \times B^m) \cap g_j(B^n \times B^m) \neq \emptyset$ (here $\rho_i : R^n \times R^m \to R^n$ is projection onto first factor). Otherwise set $M_{u,i,j} = \emptyset$. Define $M_{s,i,j}$ similarly. Then for each $i \in I$ it is required that the smooth submanifolds $\{M_{u,i,j} : j \in I\}$ are in transverse position to one another in R^n, and the smooth submanifolds $\{M_{s,i,j} : j \in I\}$ are in transverse position to one another in R^m.

Let $K_{u,i}, K_{s,i}$ denote smooth triangulations for the $M_{u,i}, M_{s,i}$. For any $i, j \in I$ with $M_{u,i,j} \neq \emptyset$ we define a smooth triangulation $K_{u,i,j}$ for $M_{u,i,j}$ as follows: $e \in K_{u,i,j}$ if and only if there is $e' \in K_{u,j}$ such that $e = \rho_i \circ g_i^{-1} \circ g_j(e' \times B^m)$. If $M_{u,i,j} = \emptyset$ then set $K_{u,i,j} = \emptyset$. Define smooth triangulations $K_{s,i,j}$ for the $M_{s,i,j}$ similarly.

The techniques of piecewise smooth triangulation theory allow us to choose the $\{K_{u,i} : i \in I\}$ and $\{K_{s,i} : i \in I\}$ so as to satisfy the following:

6.4. (a) For each $i \in I$ the $\{K_{u,i,j} : j \in I\}$ are in transverse position to one another in R^n, and the $\{K_{s,i,j} : j \in I\}$ are in transverse position to one another in R^m.

(b) For any $i \in I$ and any subset $J \subset I$ the smooth polyhedron which is generated by the $\{K_{u,i,j} : j \in J\}$ (or by the $\{K_{s,i,j} : j \in I\}$), is a regular cell structure (see figure 6.5).

(c) The diameter of any simplex in any $K_{u,i,j}$ or $K_{s,i,j}$ is much less than the distance from any $\partial M_{u,i}$ to ∂B^n or from $\partial M_{s,i}$ to ∂B^m.

Figure 6.5. (In the following figure the heavy solid lines indicate portions of the boundaries $M_{u,i,j}$ and $M_{u,i,j'}$, the light solid lines indicate a portion of $K_{u,i,j}$, and the light dotted lines indicate a portion of $K_{u,i,j'}$.)

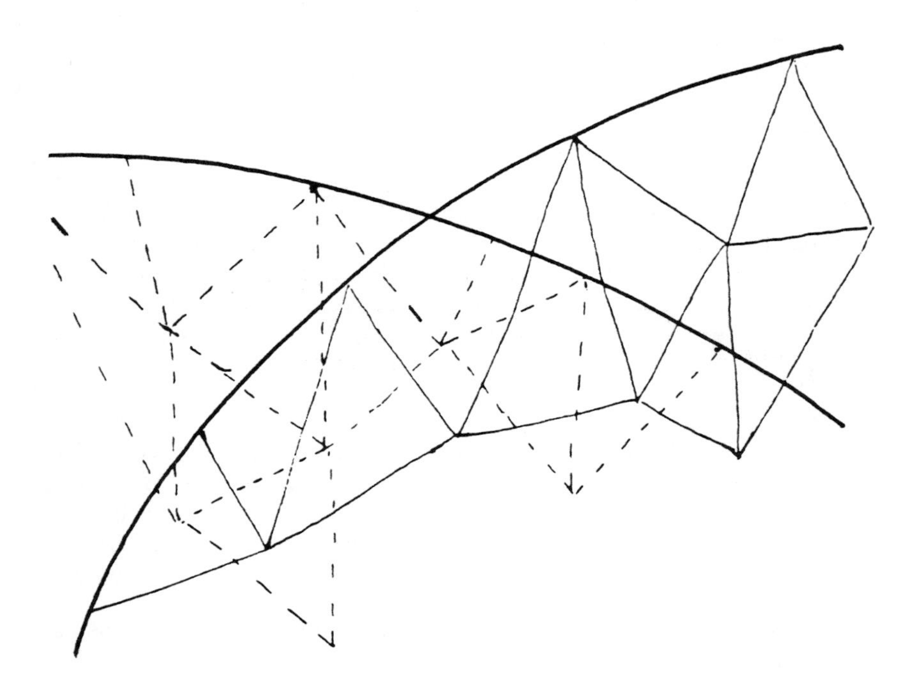

For any $i \in I$ and any $e_u \in K_{u,i}$, $e_s \in K_{s,i}$ we consider the rectangular cell $g_i(e_u \times e_s)$. There is a rectangular subdivision of $g_i(e_u \times e_s)$—denoted by $S(e_u \times e_s)$—defined as follows. Let J denote the subset of all $j \in I$ such that for some $e'_u \in K_{u,i}$, $e'_s \in K_{s,i}$ (which depend on j and satisfy $e_u \subset e'_u, e_s \subset e'_s$) we have that $g_i(e'_u \times e'_s) \cap g_j(B^n \times B^m) \neq \emptyset$. Let Y_u denote the regular cell complex generated by the $\{K_{u,i,j} : j \in J\}$, and let Y_s denote the regular cell complex generated by the $\{K_{s,i,j} : j \in J\}$ (see 6.4(b)). Denote by X the subcomplex of $Y_u \times Y_s$ satisfying $|X| = e_u \times e_s$. Set $S(e_1 \times e_2)$ equal to the image X under $g_i : B^n \times B^m \to V$.

We can now define the rectangular cell complex C of 6.2.

6.6. Define C as the cell complex with the least number of cells which satisfies the following:

(a) $|C| = \cup_{i \in I}\, g_i(M_{u,i} \times M_{s,i})$.

(b) For each $i \in I$ and each $e_u \in K_{u,i}$, $e_s \in K_{s,i}$ the rectangular cell complex $S(e_u \times e_s)$ is subdivided by C.

It is left as an exercise to show from 6.4 that C is well defined by 6.6.

If the $K_{u,i}, K_{s,i}$ are chosen so that the diameter of each triangle in any $K_{u,i}$ or $K_{s,i}$ is sufficiently small then C will satisfy 6.2(a).

Note that 6.3(a), 6.6(a) imply 6.2(b).

Finally note that the $\{K_{u,i}, K_{s,i} : i \in I\}$ can be chosen to also satisfy $\sigma \leq \tau(X)$, $\sigma \leq D(X)/D(Y)$ where X, Y stand for any of the $\{K_{u,i,j}, K_{s,i,j} : i, j \in I\}$ and σ is a positive number depending only on η of 6.1(e) and on $n+m$. Then 6.2(c) will hold, where the b in 6.2(c) depends only on $n+m, \sigma$, and η of 6.1(e).

This completes the proof of lemma 6.2.

For each cell $e \in C$ we choose one of the $\{g_i : i \in I\}$ which satisfies $e \subset g_i(B^n \times B^m)$ and denote this choice by $g_e : R^n \times R^m \to M$. There are cells $e_u \subset R^n, e_s \subset R^m$ uniquely determined by the requirement $g_e(e_u \times e_s) = e$. We choose open sets $X_{u,e}, X_{s,e}$ in R^n, R^m respectively which satisfy the following properties.

6.7. (a) For each $e \in C$ the union $\cup_{e' \in e}\, g_{e'}(X_{u,e'} \times X_{s,e'})$ is a neighborhood for e in M. This neighborhood may be assumed to be arbitrarily small for each $e \in C$.

(b) We have $g_e(\overline{X_{u,e} \times X_{s,e}}) \cap g_{e'}(\overline{X_{u,e'} \times X_{s,e'}}) \neq \emptyset$ if and only if either $e \subset e'$ or $e' \subset e$.

(c) $e \cap g_{e'}(X_{u,e'} \times X_{s,e'}) \neq \emptyset$ if and only if $e' \subset e$.

The collection $\{g_e, X_{u,e}, X_{s,e} : e \in C\}$ is referred to throughout the remainder of this paper.

Let $\{K_{u,e} : e \in C\}$ denote a collection of full smooth polyhedra in the $\{X_{u,e} : e \in C\}$. If $e \subset e'$ or $e' \subset e$ then let $K_{u,e,e'}$ denote the smooth polyhedron which is the image of $K_{u,e'}$ under the composite map

$$R^n = R^n \times 0 \xrightarrow{g_{e'}} M \xrightarrow{g_e^{-1}} R^n \times R^m \xrightarrow{\text{proj.}} R^n.$$

Otherwise set $K_{u,e,e'} = \emptyset$. For any subcomplex $A \subset C$ denote by $K(A, e)$ the structure generated by the $\{K_{u,e,e'} : e' \in A\}$.

The following proposition is an easy application of propositions 2.10 and 4.8. (We remind the reader that the terminology "almost linear polyhedron" was introduced in remark 4.11.)

PROPOSITION 6.8. *Given any $\varepsilon > 0$, and positive integers $r_0 < r_1 < \cdots < r_{n+m}$, there is a collection of almost linear polyhedra $\{K_{u,e} : e \in C\}$ in the $\{X_{u,e} : e \in C\}$ which satisfy the following properties.*

(a) For any given compact subsets $N_{u,e} \subset X_{u,e}$ it may be assumed that $N_{u,e} \subset Int(K_{u,e})$ for each $e \in C$.

(b) For any given $e \in C$ let B denote the subcomplex of all $e' \in C$ such that $e' \subset e'', e \subset e''$ for some $e'' \in C$. Let K_B denote the collection

of all $\{K_{e'} : e' \in B\}$, where $K_{e'}$ is equal $K_{u,e,e'}$. Then the collection of full smooth polyhedra K_B are in B-transverse position to one another.

(c) There is for each $i \in \{0, 1, \ldots, n+m\}$ a number $\mu_i > 0$ which depends only on $(n+m, r_i, N(C), \eta)$ and satisfies $\mu_i < \mu_{i-1}$ for all i. For any $e \in C$ and B as in (b) we have that $\mu_i < \tau(K_{B^i})$, $\mu_i < \tau(K_{u,e})$, and $\mu_i < D(K_{u,e})/D(K_{u,e'})$ for all $e, e' \in C^i$.

(d) For each $e \in C^i - C^{i-1}$ and B as in (b) there is an almost linear triangulation $L_{u,e}$ in $X_{u,e}$ and an almost linear r_i-derived subdivision $L_{u,e}^{(r_i)}$ of $L_{u,e}$. We have that $K_{u,e} = L_{u,e}^{(r_i)}$. There are subcomplexes $L'_{u,e}, K'(\partial e)$ of $L_{u,e}$, $K(\partial e)$—recall that $K(\partial e)$ is the almost linear polyhedron generated by the $\{K_{e'} : e' \in \partial e\}$—such that $|L'_{u,e}| = |K'(\partial e)| = |L_{u,e}| \cap |K(\partial e)|$ and such that $L'_{u,e}$ subdivides $K'(\partial e)$. Moreover $\mu_i < D(K_{u,e})/D(L_{u,e})$, $\mu_i < \tau(L_{u,e})$.

(e) $\overline{D}(L_{u,e}) < \varepsilon$ holds for all $e \in C$.

Proof of Proposition 6.8. Note that there is a positive integer s, and for each $i \in \{0, 1, \ldots, n+m\}$ and each $j \in \{1, 2, \ldots, s\}$ there is a subcomplex $C^{i,j} \subset C$ which satisfy the following.

6.9. (a) the integer s depends only on $N(C)$.

(b) $C^i \subset C^{i,j} \subset C^{i,j+1}$ for all i, j; and $C^{i,s} = C^{i+1}$.

(c) For any i, j and any $e, e' \in C^{i,j} - C^{i,j-1}$ there is no cell in C which contains both e and e'.

The proof of 6.8 proceeds by induction over the sequence $\ldots, C^{i,j}, C^{i,j+1}, C^{i,j+2}, \ldots$.

Suppose that all the $\{K_{u,e} : e \in C^{i,j}\}$ have been constructed so as to satisfy 6.8(a)-(e).

For any $e \in C^{i,j+1} - C^{i,j}$ we construct $K_{u,e}$ as follows. Let B denote the subcomplex of C associated to e in 6.8(b). Set $A = B \cap C^{i,j}$, and for each $e' \in A$ set $K_{e'} = K_{u,e,e'}$. Let K_A denote the collection $\{K_{e'} : e' \in A\}$. We apply propositions 4.8, 2.10 to extend the collection K_A to a collection K_B of almost linear polyhedra $\{K_{e'} : e' \in B\}$. Note by 2.10(c) we have $K_e = L_e^{(r_{i+1})}$ for some almost linear triangulation L_e (where $r = r_{i+1}$ in 2.10(c)). By 2.10(b) we may assume $|L_e| \supset \overline{X}_{u,e}$. By 2.10(c) a subcomplex of L_e subdivides $K(\partial e)$. Thus if $L_{u,e}$ is defined to be a maximal full subcomplex of L_e which satisfies 6.10 below, and $K_{u,e}$ is defined to be $L_{u,e}^{(r_{i+1})}$, then the $\{K_{u,e'} : e' \in C^{i,j+1}\}$ will satisfy 6.8(a)-(e) provided $D(L_e)$ and $\{D(K_{u,e'}) : e' \in C^{i,j}\}$ are sufficiently small.

6.10. (a) $|K(\partial e)| \cap |L_{u,e}|$ is the underlying set of a subcomplex of $K(\partial e)$.

(b) $|L_{u,e}| \subset X_{u,e}$.

§7. SMOOTH BALL STRUCTURES NEAR Λ

We use the term **piecewise smooth** (P.S.) **cell** to mean any smooth polyhedron Y in R^n which has a linear parametrization (h, L) such that $|L| = \{x_1, \ldots, x_r, 0, \ldots, 0\} \in R^n : 0 \leq x_i \leq 1\}$ for some $r \in \{1, 2, \ldots, n\}$. A **P.S. ball** in R^n is an n-dimensional P.S. cell in R^n.

DEFINITION 7.0. *Let $K_{u,e}$ be as in 6.8 for some $e \in C$. A P.S. ball structure for $K_{u,e}$ consists of a collection of P.S. balls $\{Y(f) : f \in K_{u,e}\}$ in $X_{u,e}$—one such ball for each $f \in K_{u,e}$—which satisfy the following.*

(a) For any subcomplex $L \subset K_{u,e}$ we have that $\cup_{f \in L} Y(f)$ is a neighborhood for L in R^n.

(b) For any $f, f' \in K_{u,e}$ we have that $Y(f) \cap f' \neq \emptyset$ if and only if $f \subset f'$, and $Y(f) \cap Y(f') \neq \emptyset$ if and only if either $f \subset f'$ or $f' \subset f$.

DEFINITION 7.1. *A* **redundant P.S. ball structure** *for $K_{u,e}$ consists of a finite set $\{Y_i(f) : f \in K_{u,e}\}$, $i = 1, 2, \ldots, x$, of P.S. ball structures for $K_{u,e}$ which satisfy the following:*

(a) For any $i, j \in \{1, 2, \ldots, x\}$ and any $f \in K_{u,e}$ we must have either $Y_i(f) \subset \mathrm{Int}(Y_j(f))$ or $Y_j(f) \subset \mathrm{Int}(Y_i(f))$

(b) For each $f \in K_{u,e}$ randomly choose one of the $\{Y_i(f) : 1 \leq i \leq x\}$ and denote it by $Y(f)$. Then the $\{Y(f) : f \in K_{u,e}\}$ must be a P.S. ball structure for $K_{u,e}$.

For any $e, e' \in C$ and a redundant P.S. ball structure $\{Y_i(f) : f \in K_{u,e'}, 1 \leq i \leq x\}$ for $K_{u,e'}$ we define a redundant P.S. ball structure $\{Y_{i,e}(f') : f' \in K_{u,e,e'}, 1 \leq i \leq x\}$ for $K_{u,e,e'}$ as follows. If $e \subset e'$ or $e' \subset e$ then let $Y_{i,e}(f'), f'$ be the images of $Y_i(f), f$ under the composite map

$$R^n = R^n \times 0 \xrightarrow{g_{e'}} M \xrightarrow{g_e^{-1}} R^n \times R^m \xrightarrow{\text{proj.}} R^n.$$

Otherwise set $Y_{i,e}(f') = \emptyset = f'$.

LEMMA 7.2. *Given a positive integer x there are redundant P.S. ball structures $\{Y_i(f) : f \in K_{u,e}, 1 \leq i \leq x\}$ for each $\{K_{u,e} : e \in C\}$ such that the following properties hold.*

(a) For each $e \in C$ all the smooth polyhedra $\{Y_{i,e}(f) : f \in K_{u,e,e'}, e' \in C, 1 \leq i \leq x\} \cup \{K_{u,e,e'} : e' \in C\}$ are actually almost linear polyhedra. Let $V_{u,e}$ denote the almost linear polyhedron generated by the $\{K_{u,e,e'} : e' \in C\}$. All the $\{Y_{i,e}(f) : f \in K_{u,e,e'} : e' \in C, 1 \leq i \leq x\}$ are in transverse position to one another and generate an almost linear polyhedron denoted by $W_{u,e}$. $V_{u,e}$ and $W_{u,e}$ are in transverse position to one another and generate an almost linear polyhedron $T_{u,e}$.

(b) There is a number $\mu \in (0,1)$ which depends only on $(x, n, N(C))$ and on the number μ_{n+m} of 6.8(c). We have that μ is a lower bound for all of the $\{\tau(X_{u,e}), D(X_{u,e})/D(K_{u,e'}) : e, e' \in C\}$, where $X = V, W$, or T.

(c) Let $e_1, e_2, \ldots, e_k \in C$ be such that $e_i \subset e_{i+1}$ for all $i \in \{1, 2, \ldots, k-1\}$. Let e denote any one of the $e_1, \ldots, e_k$. Suppose $f \in K_{u,e,e_i}$, $f' \in K_{u,e,e_{i'}}$ with $i > i'$. If $f \subset f' - \partial f'$ we require that $Y_{j,e}(f) \subset \text{Int}(Y_{j',e}(f'))$ for all $j', j \in \{1, 2, \ldots, x\}$. Moreover, if $f \subset f'$ (but $f \not\subset f' - \partial f'$) then we require that $\cup_{g \in f} Y(g) \subset \text{Int}(\cup_{g' \in f'} Y(g'))$, where $Y(g), Y(g')$ can be any of $\{Y_{j,e}(g), Y_{j,e}(g') : 1 \le j \le x\}$.

(d) Let $e, e_1, e_2, \ldots, e_k$ be as in (c). Let J be a subset of $\cup_{i=1}^k K_{u,e,e_i}$ satisfying: for any $f_1, f_2 \in J$ if $f_1 \subset f_2$ and $f_1 \not\subset \partial f_2$ then $f_1 = f_2$. Then for $J' \subset J$ and any $g \in \cup_{i=1}^k K_{u,e,e_i}$ we require that

$$g \bigcap \left(\bigcap_{f \in J - J'} Y(f) \right) \bigcap \left(\bigcap_{f' \in J'} \partial Y(f') \right) \neq \emptyset$$

if and only if the following three properties hold. (Here $Y(f)$ denotes any of the $\{Y_{j,e}(f) : 1 \le j \le x\}$).

(i) There is an ordering $h_1, h_2, \ldots, h_l$ of all the simplices in J. For each $1 \le i < j \le l$ we have $(h_i - \partial h_i) \cap h_j \neq \emptyset$ and $h_i \cap h_j \subset \partial h_j$. Also $e_{i'} \subset e_{j'}$, where $h_i \in K_{u,e,e_{i'}}$ and $h_j \in K_{u,e,e_{j'}}$.

(ii) $g \cap (h_i - \partial h_i) \neq \emptyset$ for all i.

(iii) $\dim(g) - |J'| \ge 0$, where $|J'|$ denotes the order of J'.

(e) Suppose that J, J' are as in (d) and that if $J - J' = \emptyset$ then $g \notin J'$. Then any non-empty intersection $g \cap (\cap_{f \in J-J'} Y(f)) \cap (\cap_{f \in J'} \partial Y(f))$ of (d) is a P.S. cell of dimension equal to $\dim(g) - |J'|$, where $|J'|$ is the order of J'. Moreover if $J - J' \neq \emptyset$ then any non-empty intersection $\partial g \cap (\cap_{f \in J-J'} Y(f)) \cap (\cap_{f \in J'} \partial Y(f))$ is a P.S. cell of dimension equal $\dim(g) - |J'| - 1$. Finally, if $J - J' \neq \emptyset$ then $(\cap_{f \in J-J'} Y(f)) \cap (\cap_{f \in J'} \partial Y(f))$ is a P.S. cell of dimension equal $n - |J'|$.

PROOF OF LEMMA 7.2: We shall first complete the proof of 7.2 under the following assumption.

7.4. (a) All the $\{K_{u,e} : e \in C\}$ are linear triangulations in R^n.
(b) Every composite map

$$R^n = R^n \times 0 \subset R^n \times R^m \xrightarrow{g_e} M \xrightarrow{g_{e'}^{-1}} R^n \times R^m \xrightarrow{\text{proj.}} R^n$$

is a locally linear map wherever it is defined.

For each $e \in C$, $f \in K_{u,e}$, and $i \in \{1, 2, \ldots, x\}$, the P.L. ball $Y_i(f)$ will have the following form. Let P denote the plane in R^n generated by f, and let P_j, $j = 1, 2, \ldots, \dim(f) + 1$, denote the codimension one planes in P which are generated by the codimension one faces of f. Set b_j equal to the barycenter of the face of f which generates P_j and let v_j be the unit vector in P which is normal to P_j and points into f at b_j. Define mappings $r_j : P \to R$ by $r_j(p) = \langle v_j, p - b_j \rangle$. Note that f equals the intersection $\cap_{i=1}^{\dim(f)+1} r_i^{-1}((0, \infty))$. Let $P^\perp$ denote the plane in R^n of dimension n-$\dim(f)$ which intersects P perpendicularly at the barycenter $b \in f$ of the simplex f. We require that each $Y_i(f)$ have the following form.

7.5. $Y_i(f) = \cup_{p \in \hat{f}} \Delta(p)$. Here $\hat{f} = \cap_{j=1}^{\dim(f)+1} r_j^{-1}([\varepsilon_j, \infty))$ for some positive numbers ε_j, $i = 1, 2, \ldots, \dim(f) + 1$, Δ is a linear simplex in $P^\perp$ of dimension equal $n - \dim(f)$ such that $f \cap (\Delta - \partial\Delta) = b$, and for each $p \in f$ $\Delta(p)$ denotes the image of Δ under translation by $p - b$.

In the rest of this proof we always assume that each $Y_i(f)$ is as in 7.5. Subject to these restrictions we are still free to vary the ε_j, $j = 1, 2, \ldots, \dim(f) + 1$, and Δ of 7.5.

The construction of the $\{Y_i(f)\}$ will proceed by induction over the dimensions of cells $e \in C$ and over the dimensions of simplices $f \in K_{u,e}$. Here is the induction hypothesis. Set $S_{k,\ell} = \{f : f \in K_{u,e}$ and $\dim(e) < k$, or $f \in K_{u,e}$ with $\dim(e) = k$ and $\dim(f) \leq \ell\}$. For any $e \in S$ set $S_{k,\ell,e}$ equal the images of all the $f \in S_{k,\ell}$ under the composite maps

$$R^n = R^n \times 0 \xrightarrow{g_{e'}} M \xrightarrow{g_e^{-1}} R^n \times R^m \xrightarrow{\text{proj.}} R^n,$$

where $f \in K_{u,e'}$ and either $e \subset e'$ or $e' \subset e$.

HYPOTHESIS 7.6(k, ℓ): All the $\{Y_i(f) : f \in S_{k,\ell}, 1 \leq i \leq x\}$ have been constructed and satisfy the following properties.

(a) If $\dim(e) < k$ then the $\{Y_i(f) : f \in K_{u,e}, 1 \leq i \leq x\}$ are a redundant P.L. ball structure for $K_{u,e}$. If $\dim(e) = k$ then the $\{Y_i(f) : f \in K_{u,e}^\ell, 1 \leq i \leq x\}$ are a redundant P.L. ball structure for $K_{u,e}^\ell$.

(b) For any $e \in C$ the linear polyhedra $\{Y_{i,e}(f) : f \in S_{k,\ell,e}, 1 \leq i \leq x\}$ are in transverse position to one another and generate a linear polyhedron denoted by $W_{u,e}^{k,\ell}$. Let $V_{u,e}$ denote the linear polyhedron generated by all the $\{K_{u,e,e'} : e' \in C\}$. The $W_{u,e}^{k,\ell}$, $V_{u,e}$ are in transverse position to one another.

(c) Suppose in 7.2(c)(d) we have that $J \subset S_{k,\ell,e}$. Then 7.2(c)(d)(e) hold for any $J' \subset J$ and any $g \in \cup_{i=1}^k K_{u,e,e_i}$.

The induction step consists of showing that $7.6(k,\ell) \Rightarrow 7.6(k,\ell+1)$ (or of showing $7.6(k,\ell) \Rightarrow 7.6(k+1,0)$ if $\ell = n$). The induction step is broken into the following two steps.

STEP I. In this step we construct, for each $f \in S_{k,\ell+1} - S_{k,\ell}$ the set $\hat{f}$ of 7.5 so that $\hat{f}$ satisfies the following.

7.7. (a) Let $e \in C$ be such that $f \in K_{u,e}$. For any $e' \in C$, with either $e' \subset e$ or $e \subset e'$, let $f' \in K_{u,e',e}$ be the simplex corresponding to f. Let A be the linear polyhedron in P (where P is the plane spanned by f') whose members have the form $f' \cap d \cap d'$, where $d \in W^{k,\ell}_{u,e'}$, $d' \in V_{u,e'}$. Let $\hat{f}' \subset f'$ correspond to $\hat{f} \subset f$. Then we require that the simplex $\hat{f}'$ and the linear polyhedron A be in transverse position in P.

(b) For any $e' \in C$, $f' \in K_{u,e,e'}$ satisfying $f' \subset f$, and for any subsets $J \subset S_{k,\ell,e}$, $J' \subset J$ we must have that the intersection

$$(\cap_{g\in J'}\partial Y(g)) \cap (\cap_{g\in J-J'}Y(g)) \cap f' \cap \hat{f}$$

and

$$(\cap_{g\in J'}\partial Y(g)) \cap (\cap_{g\in J-J'}Y(g)) \cap \partial(f' \cap \hat{f})$$

are P.L. homeomorphic to the intersections

$$(\cap_{g\in J'}\partial Y(g)) \cap (\cap_{g\in J-J'}Y(g)) \cap f'$$

and

$$(\cap_{g\in J'}\partial Y(g)) \cap (\cap_{g\in J-J'}Y(g)) \cap \partial f'.$$

For e', f' as 7.7(a) let A_1 equal the linear polyhedron whose members are of the form $d \cap d'$ where $d \in f'$, $d' \in W^{k,\ell}_{u,e'}$, and set A_2 equal the P.L. polyhedron whose members are of the form $f' \cap d$ where $d \in V_{u,e'}$. Note that A_1, A_2 generate A of 7.7(a), and that A_1, A_2 are in transverse position modulo $\partial f'$ in the plane P. Next note that because the polyhedra $f', W^{k,\ell}_{u,e'}$ are in transverse position (see 7.6(k,ℓ)(b)) there is $\delta > 0$ satisfying the following.

7.8. If $\varepsilon_j \leq \delta$ holds for all j in 7.5 then the linear polyhedra $\hat{f}', A_1$ are in transverse position in the plane P, where $\hat{f}'$ is as in 7.7(a).

Now note for almost all $(\dim(f)+1)$-tuples $(\varepsilon_1, \varepsilon_2, \ldots, \varepsilon_{\dim(f)+1})$ in $[0,\delta]^{\dim(f)+1}$ 7.7(a) is true for the corresponding $\hat{f}$.

Now we let e', f' be as in 7.7(b). To verify 7.7(b) we need a special vector field on $\partial f'$. Note that because f' is in transverse position to $W^{k,\ell}_{u,e}$ (see $7.6(k,\ell)$(b)) there is a vector field $\psi : \partial f' \to T(P')$ satisfying the following properties, where P' is the plane in R^n generated by f'.

7.9. (a) For each $q \in \partial f'$ the vector $\psi(q)$ points outward in P' from f'.

(b) If for any $d \in W^{k,l}_{u,e}$ we have that $q \in d \cap \partial f'$ then $\psi(q)$ must be tangent to d.

(c) There is $\varepsilon > 0$ such that for any codimension one face $\partial_i f'$ of f' and any $q \in \partial_i f'$ the angle between $\psi(q)$ and $T(\partial_i f')$ is greater than ε.

(d) For each $q \in \partial f$ we have $\frac{1}{2} \leq |\psi(q)| \leq 2$. Moreover $\psi : \partial f' \to T(P')$ is a P.L. mapping.

We consider now the mapping $r : \partial f' \times R \to P'$ defined by $r(q,t) = q + t\psi(q)$. Note it follows from 7.9(c)(d) that for sufficiently small $\varepsilon' > 0$ the following holds.

7.10. $r : \partial f' \times [-\varepsilon', \varepsilon'] \to P'$ is a well defined P.L. collaring for $\partial f'$ in P'.

Now we can complete the verification of 7.7(b). Recall that we are choosing $\hat{f}$ as in 7.5 with the ε_j of 7.5 satisfying $\varepsilon_j \leq \delta$ for all j, where δ comes from 7.8. We need to take δ sufficiently small so that in addition to satisfying 7.8 δ also satisfies the following property.

7.11. If $\varepsilon_j \leq \delta$ holds for all j in 7.5 then we must have that $f' - (f' \cap \hat{f}) \subset r(\partial f' \times [-\varepsilon'/2, \varepsilon'/2])$, where r comes from 7.10 and f, f' are any simplices with $f \in S_{k,l+1} - S_{k,l}$ and f' is as in 7.7(b).

Note that 7.11 makes possible the following definition of a P.L. homeomorphism $h : P' \to P'$.

7.12 $h(p) = p$ if $p \in P' - r(\partial f' \times [-\varepsilon', \varepsilon'])$; if $p = r(q,t)$ for $(q,t) \in \partial f' \times [-\varepsilon', \varepsilon']$ then

$$h(p) = \begin{cases} r(q, t(\frac{\varepsilon' + \tau(q)}{\varepsilon'}) + \tau(q)) \text{ if } t \in [-\varepsilon', 0] \\ r(q, t(\frac{\varepsilon' - \tau(q)}{\varepsilon'}) + \tau(q)) \text{ if } t \in [0, \varepsilon']. \end{cases}$$

Here $\tau(q) \in [-\varepsilon', \varepsilon']$ is defined by the requirement that $r(q, \tau(q)) \in \partial(\hat{f} \cap f')$. [Note that for δ sufficiently small in 7.8 $\tau(q)$ is well defined by this requirement.]

It follows from 7.12, 7.8, 7.9(b) that for each $d \in W^{k,\ell}_{u,e}$ $h : P' \to P'$ maps $|d| \cap |f'|$ onto $|d| \cap |\hat{f} \cap f'|$. Thus any union $\cup_{d \in W} d \cap f'$ is homeomorphic to $\cup_{d \in W} d \cap (\hat{f} \cap f')$ via $h : P' \to P'$, for any subset $W \subset W^{k,\ell}_{u,e}$. Now note that the sets $X \cap f'$ and $X \cap (\hat{f} \cap f')$ are of the form $\cup_{d \in W} d \cap f'$ and $\cup_{d \in W} d \cap (\hat{f} \cap f')$ where $X = (\cap_{g \in J'} \partial Y(g)) \cap (\cap_{g \in J - J'} Y(g))$ comes from 7.7(b).

STEP II. In this step we complete the argument for the induction step 7.6(k,ℓ) $\Rightarrow$ 7.6(k,$\ell+1$).

Let $e, e' \in C$, and suppose $f_1 \in K_{u,e}$, $f_i \in K_{u,e,e'}$ for $i = 2,3$, satisfy the following: $f_1 \in S_{k,l+1} - S_{k,l}$; $f_1 \cap f_2 = f_3$. For any two simplices $f, f' \in f_2$ we denote by $f \star f'$ the simplex of f_2 generated by the vertices of f and f'. Let $f_4 \in f_2$ denote the simplex satisfying $f_3 \star f_4 = f_2$, $f_4 \cap f_3 = \emptyset$. For any simplex f in R^n let $P(f)$ denote the plane generated by f. Choose a vector bundle $\xi(f_1, f_2)$ over f_3 which satisfies the following properties (see 7.6(k,ℓ)(b) and 7.7(a)).

7.13. (a) $\xi(f_1, f_2)$ is a subbundle of $T(P(f_2))|f_3$ and has fiber of dimension equal $\dim(f_2) - \dim(f_3)$.

(b) For any simplex $f \in f_3$ and any point $q \in f$ we must have that $\xi(f_1, f_2)_q$ is tangent to $P(f \star f_4)$.

(c) For any $d \in W^{k,\ell}_{u,e}$ and any point $q \in d \cap f_3$ we must have that $\xi(f_1, f_2)_q$ is tangent to $P(d)$, where $P(d)$ is the plane generated by d.

(d) The transition functions for the bundle $\xi(f_1, f_2)$ are piecewise smooth maps.

(e) There exists an $\varepsilon > 0$ such that for all points $q \in f_3$ the angle between $\xi(f_1, f_2)_q$ and $P(f_3)$ is greater than ε.

(f) For any face d of the simplex $\hat{f}_1$ let $Q(d)$ denote the union $\cup_{x \in P(d)} P^{\perp}_x$, where $P^{\perp}_x$ denotes the maximal plane perpendicular to $P(f_1)$ at x. Then for each $q \in d \cap f_3$ we must have that $\xi(f_1, f_2)_q$ is tangent to $Q(d)$.

Let exp: $\xi(f_1, f_2) \to P(f_2)$ denote the exponential map, and for any $\delta > 0$ we let $\xi_\delta(f_1, f_2)$ denote the subset of all $p \in \xi(f_1, f_2)$ such that $\exp(p)$ is a distance less than or equal to δ from $P(f_3)$. Note that it follows from 7.13(a)(d)(e) that there is an $\varepsilon' > 0$ such that the following is satisfied for all the $\xi(f_1, f_2)$ of 7.13.

7.14. exp: $\xi_{\varepsilon'}(f_1, f_2) \to P(f_2)$ is a P.L. embedding.

It follows from 7.7(a) that there is $\varepsilon'' > 0$ sufficiently small so that the following are satisfied.

7.15. (a) For any $f \in S_{k,\ell+1}$ let $Y_i(f)$, $\hat{f}$, and Δ be as in 7.5. Then if diameter$(\Delta) \leq \varepsilon''$ we have that $Y_i(f) \cap d \neq \emptyset$ if and only if $\hat{f} \cap d \neq \emptyset$. Here $f \in K_{u,e}$ and $d \in W^{k,\ell}_{u,e} \cup V_{u,e}$.

(b) Suppose (f_1, f_2, f_3) are as in 7.13 with $f_1 = f$, and $Y_i(f)$ is as in (a). Then we must have that

$$f_2 \cap Y_i(f) \subset \exp(\xi_{\varepsilon'}(f_1, f_2)) \cap f_2.$$

Finally note the following property. For any $f \in S_{k,\ell+1} - S_{k,\ell}$ let Δ, $P^{\perp}, b$ be as in 7.5, let $\{q_i : 1 \leq i \leq y\}$ denote the vertices of Δ, and let $B_{\varepsilon''}$ denote the ball of radius ε'' centered at b in $P^{\perp}$. (The f of 7.5 is now equal to f of 7.15.)

7.16. For almost all y-tuples $(q_1, q_2, \ldots, q_y)$ in $(B_{\varepsilon''} - b)^y$ the $Y_i(f)$ constructed from the simplex $\Delta = \langle q_1, q_2, \ldots, q_y \rangle$ in 7.5 will satisfy 7.6$(k, \ell+1)$(b). (Compare with 7.7(a).)

For each $f \in S_{k,\ell+1} - S_{k,\ell}$ we choose the $\{Y_i(f) : 1 \leq i \leq x\}$ of 7.5 to satisfy 7.15, 7.16. It follows from 7.6(k,ℓ), 7.15, 7.16, 7.7(b) that the $\{Y_i(f) : f \in S_{k,\ell+1},\ 1 \leq i \leq x\}$ satisfy 7.6$(k, \ell+1)(a)(b)$. So to complete the induction step 7.6$(k,\ell) \Rightarrow$ 7.6$(k, \ell+1)$ it remains to verify 7.6$(k,\ell+1)(c)$.

Note that the portion of 7.6$(k, \ell+1)(c)$ referring to 7.2$(c)(d)$ is an immediate consequence of 7.5, 7.15, 7.7(b), and 7.6(k,ℓ).

We now verify the portion of 7.6(k,ℓ+1)(c) which refers to 7.2(e). We begin by giving to the set $Y_i(f_1) \cap f_2$ a bundle structure over $f_3 \cap \hat{f}_1$ denoted by $\gamma(f_1, f_2)$—where (f_1, f_2, f_3) is any triple as in 7.13. For each $q \in f_3 \cap \hat{f}_1$ the fiber $\gamma(f_1, f_2)_q$ is defined to be $\xi(f_1, f_2)_q \cap Y_i(f_1) \cap f_2$. (Here we have identified $\xi(f_1, f_2)_q$ with its image under the map $\exp : \xi(f_1, f_2) \to R^n$.) The local trivializations for $\gamma(f_1, f_2)$ come from a global trivialization defined as follows. For any $q \in f_3 \cap \hat{f}_1$ define $\psi_q : \gamma(f_1, f_2) \to (f_3 \cap \hat{f}_1) \times \gamma(f_1, f_2)_q$ by $\psi_q(p) = (p_1, p_2)$ where $p \in \gamma(f_1, f_2)_{p_1}$ and $(P(f_3) + p - q) \cap \gamma(f_1, f_2)_q = p_2$, and $P(f_3) + p - q$ is the image of $P(f_3)$ under translation by $p - q$. Note it follows from 7.13, 7.14, 7.15 that ψ_q is a well defined homeomorphism mapping each fiber $\gamma(f_1, f_2)_{p_1}$ onto $p_1 \times \gamma(f_1, f_2)_q$. In the notation of 7.2(e) we must show that any intersection

$$Y = \left(\bigcap_{f \in J'} \partial Y(f) \right) \bigcap \left(\bigcap_{f \in J - J'} Y(f) \right) \bigcap g$$

is either empty or a P.L. cell of dimension equal to $\dim(g) - |J'|$ where $J \subset S_{k,l+1}$, $J' \subset J$, and $g = f_2$. Moreover if $J - J' \neq \emptyset$ we must show that

$$Y \cap \partial g$$

is either empty or a P.L. cell of dimension equal $\dim(g) - |J'| - 1$. If Y (or $Y \cap \partial g$) is non-empty then we must have that $J \cap (S_{k,\ell+1,e} - S_{k,\ell,e}) = J''$ is either the empty set or contains just one simplex (which we denote by f_1 in anticipation of using the bundle $\gamma(f_1, f_2)$). If $J'' = \emptyset$ then 7.6(k,ℓ) applies to Y (and $Y \cap \partial g$) to show they are either empty sets or P.L. cells of the desired dimensions. If $J'' = \{f_1\}$ there are the following four cases to consider.

REMARK. In the four following cases we assume that $f_1 \in K_{u,e}$ and $g \in K_{u,e,e'}$, where $\dim(e) \leq \dim(e')$. (This is the case for $f_1, g = f_2$ of 7.13.) The case when $\dim(e') < \dim(e)$ is similar to cases 1, 2 below, so is left as an exercise.

CASE 1. Suppose $f_1 \in J - J'$ and $f_1 \cap g = f_1$. Then Y is equal to the restriction of the bundle $\gamma(f_1, f_2)$ to the subset

$$X = \left(\bigcap_{f \in J'} \partial Y(f)\right) \bigcap \left(\bigcap_{f \in J - J' - \{f_1\}} Y(f)\right) \bigcap \left(f_3 \bigcap \hat{f}_1\right).$$

Note that X is either the empty set or a P.L. ball (see 7.7(b)). Note that the fiber $\gamma(f_1, f_2)_q$ is the intersection of the plane $\xi(f_1, f_2)_q$ with the P.L. convex cell $f_2 \cap Y(f_1)$ and therefore is a P.L. cell. It follows that Y must be a P.L. cell or the empty set. It is left to the reader to check that Y has the desired dimension.

Now we show that $Y \cap \partial g$ is either empty or is a P.L. cell of dimension equal $\dim(g) - J' - 1$ (assuming that $f_1 \in J - J'$). We have, because $f_1 \cap g = f_1$, that the following hold.

7.17. (a) $g' = f_1 \cap g$ is a simplex in g . (Here g is considered a linear triangulation containing itself and all its edges.)

(b) For any simplex $d \in g$ we have $Y \cap d \neq \emptyset$ if and only if $g' \subset d$. (This assumes $Y \cap g \neq \emptyset$.)

If $g' = g$ then $Y \cap \partial g = \emptyset$ (by 7.17). If g' is a codimension face of g then $Y \cap \partial g = Y \cap g'$ (see 7.17), and $Y \cap g' = X$ where

$$X = \left(\bigcap_{f \in J - J' - \{f_1\}} Y(f)\right) \bigcap \left(\bigcap_{f' \in J'} \partial Y(f')\right) \bigcap \left(f_3 \bigcap \hat{f}_1\right)$$

has already been shown to be a P.L. cell of the desired dimension. Now suppose $\dim(g') \leq \dim(g) - 2$. We will show that for each $d \in g$ with $g' \subset d$, $g' \neq d$, that the sets $Y \cap d$, $Y \cap \partial d$ are P.L. cells of dimension equal $\dim(d) - |J'|$, $\dim(d) - |J'| - 1$. To show this we use an induction argument over $\dim(d)$. Suppose the desired result holds for any $d' \in g$ with $g' \subset d'$, $g' \neq d'$, and $\dim(d') \leq r$. Choose $d \in g$ with $g' \subset d$, $g' \neq d$, and $\dim(d) = r + 1$. Note the $Y \cap d$ has the form

$$\left(\bigcap_{f \in J - J'} Y(f)\right) \bigcap \left(\bigcap_{f' \in J'} \partial Y(f')\right) \bigcap d,$$

and as such has already been seen to be a P.L. cell of dimension equal $\dim(d)-|J'|$. $Y \cap \partial d$ is a non-empty (see 7.17) P.L. manifold of dimension equal $\dim(d) - |J'| - 1$ (see 7.6(k,$\ell + 1$)(b)). So it will suffice to show that $Y \cap \partial d$ can be P.L. collapsed to a point (see p. 88 in [12]). To get such a collapsing we proceed by induction downward over the simplices of ∂d. For each $d' \in d^r - d^{r-1}$ we have that $Y \cap d' \searrow Y \cap \partial d'$ (i.e., $Y \cap d'$ collapses to the subset $Y \cap \partial d'$), because $Y \cap d'$ is known to be a P.L. cell and $Y \cap \partial d'$ is known to be a P.L. cell in $\partial(Y \cap d')$. Taking the union of all such collapsings we get $Y \cap d^r \searrow Y \cap d^{r-1}$ (a P.L. collapsing of $Y \cap d^r = Y \cap \partial d$ to the subset $Y \cap d^{r-1}$). In the same way we get P.L. collapsings $Y \cap d^{r-1} \searrow Y \cap d^{r-2}$, $Y \cap d^{r-2} \searrow Y \cap d^{r-3}, \ldots, Y \cap d^{r-x+1} \searrow Y \cap d^{r-x}$, where $r - x = \dim(g')$. Note that $Y \cap d^{r-x} = Y \cap g'$ (see 7.17), and $Y \cap g'$ equals the P.L. cell X. So there is a final P.L. collapsing $Y \cap d^{r-x} \searrow$ point. By composing all the preceding P.L. collapsings we get $Y \cap \partial d \searrow$ point, the desired P.L. collapsing of $Y \cap \partial d$ to a point.

CASE 2. Suppose $f_1 \in J'$ and $f_1 \cap g = f_1$. Let us first consider the case when $J - J' = \emptyset$. Set

$$X = \left(\bigcap_{f' \in J' - \{f_1\}} \partial Y(f') \right) \bigcap g.$$

Note that $Y = X \cap \partial Y(f_1)$. So in order to show that Y is a P.L. cell of dimension equal $\dim(g) - |J'|$ it will suffice to show that $X \cap Y(f_1)$ and $\overline{\partial(X \cap Y(f_1)) - (X \cap \partial Y(f_1))}$ are P.L. cells with dimensions equal $\dim(g) - |J'| + 1$ and $\dim(g) - |J'|$. Note that Case 1 applies to show that $X \cap Y(f_1)$ is a P.L. cell of the desired dimension. Case 1 also applies to show that

$$Y' = Y(f_1) \bigcap \left(\bigcap_{f' \in J' - \{f_1\}} \partial Y(f') \right) \bigcap \partial g$$

is a P.L. cell of dimension equal $\dim(g) - |J'|$. Finally note that $\overline{\partial(X \cap Y(f_1)) - (X \cap \partial Y(f_1))}$ equals Y' (here we use $J - J' = \emptyset$).

If $J - J' \neq \emptyset$ we use a different argument to show that Y is a P.L. cell, similar to the first argument in Case 1. Recall (from Case 1) that $X \cap Y(f_1)$ is the total space of the restriction of the fiber bundle $\gamma(f_1, f_2)$ to the P.L. cell

$$c = \left(\bigcap_{f' \in J' - \{f_1\}} \partial Y(f') \right) \bigcap \left(\bigcap_{f \in J - J'} Y(f) \right) \bigcap (g \cap \hat{f}_1).$$

Set $c' = c \cap \partial \hat{f}_1$. Note that

$$X \cap \partial Y(f_1) = (\dot{\gamma}(f_1, f_2)_{|c}) \cup (\gamma(f_1, f_2)_{|c'}),$$

where $\dot{\gamma}(f_1, f_2)$ is the subbundle of $\gamma(f_1, f_2)$ with fibers $\dot{\gamma}(f_1, f_2)_q = \xi(f_1, f_2)_q \cap \dot{Y}(f_1) \cap g$ and where $\dot{Y}(f_1)$ is composed of all translates of $\partial\Delta$ over $\hat{f}_1$, instead of being composed of translates of Δ over $\hat{f}_1$ as is $Y(f_1)$ (see 7.5 for Δ). Note that the fibers of $\dot{\gamma}(f_1, f_2)$ are P.L. cells. Thus $\dot{\gamma}(f_1, f_2)_{|c}$ must be a P.L. cell. Note also that it follows from 7.7(b) and 7.6(k,ℓ) that c' is a P.L. cell. Since the fibers of $\gamma(f_1, f_2)$ are P.L. cells (see Case 1) it follows that $\gamma(f_1, f_2)_{|c'}$ and $\dot{\gamma}(f_1, f_2)_{|c'}$ are both P.L. cells. The cells have the following dimensions: $\dim(\dot{\gamma}(f_1, f_2)_{|c}) = \dim(g) - |J'|$; $\dim(\gamma(f_1, f_2)_{|c'}) = \dim(g) - |J'|$; $\dim(\dot{\gamma}(f_1, f_2)_{|c'}) = \dim(g) - |J'| - 1$. It follows that $X \cap \partial Y(f_1)$ is a P.L. cell of dimension equal $\dim(g) - |J'|$. Finally we recall that $Y = X \cap \partial Y(f_1)$.

To show that $Y \cap \partial g$ is a P.L. cell of the desired dimension we use a collapsing argument almost identical to the one given in Case 1. The details are left to the reader.

This completes Case 2.

CASE 3. Suppose that $f_1 \in J - J'$ and that $f_1 \cap g \neq f_1$. The proof that $Y \cap g$ is a P.L. cell of dimension equal $\dim(g) - |J'|$ is the same as that given in Case 1.

Now we consider $Y \cap \partial g$ when $f_1 \in J - J'$ and $f_1 \cap g \neq f_1$. In this case 7.17(b) does not hold. However there is the following property which holds for such f_1, and is an analogue of 7.17. Let $h_1, h_2, \ldots, h_l$ be an ordering for the simplices in J. Let G denote the collection of all simplices $g' \in g$ satisfying: $g' \cap (h_i - \partial h_i) \neq \emptyset$ for all $1 \leq i \leq l$; each vertex of g' is contained in some $h_i - \partial h_i$ (i depends on the vertex).

7.18. (a) Each $g' \in G$ is contained in $f_1 \cap g$. (Compare with 7.2(d).)

(b) For any simplex $d \in g$ we have that $d \cap Y \neq \emptyset$ if and only if there is $g' \in G$ with $g' \subset d$.

We show that for any simplex $d \in g$, which satisfies $d \notin G$ but for which there is $g' \in G$ such that $g' \subset d$, that the sets $Y \cap d$ and $Y \cap \partial d$ are P.L. cells of dimension equal $\dim(d) - |J'|$ and $\dim(d) - |J'| - 1$. To show this we use an induction argument over $\dim(d)$. Suppose the desired result holds for all $d' \in g$ which satisfy $d' \notin G$, $g' \subset d'$ for some $g' \in G$, $\dim(d') \leq r$. Now choose $d \in g$ satisfying $d \notin G$, $g' \subset d$ for some $g' \in G$, $\dim(d) = r + 1$. Note that $Y \cap d$ has the form

$$\left(\bigcap_{f \in J - J'} Y(f)\right) \cap \left(\bigcap_{f' \in J'} \partial Y(f')\right) \cap d,$$

and as such has already been seen to be a P.L. cell with dimension equal to $\dim(d) - |J'|$. $Y \cap \partial d$ is a non-empty (see 7.18(b)) P.L. manifold of dimension equal $\dim(d) - |J'| - 1$ (see 7.6(k,$l+1$)(b)). So it will suffice to show that $Y \cap \partial d$ can be P.L. collapsed to a point. To get such a collapsing we proceed by induction (as in Case 1) downward over the simplices of ∂d to get P.L. collapsings

$$\begin{gathered} Y \cap d^r \searrow Y \cap (d^{r-1} \cup g'), \\ Y \cap (d^{r-1} \cup g') \searrow Y \cap (d^{r-2} \cup g'), \\ \vdots \\ Y \cap (d^{\ell+1} \cup g') \searrow Y \cap g'), \end{gathered}$$

where now g' denotes the maximal simplex in d which is a member of G. [The details of the construction of these collapsings are left to the reader.] By composing all these P.L. collapsings we get $Y \cap d^r \searrow Y \cap g'$. We already know that $Y \cap g'$ is a P.L. cell, so $Y \cap g' \searrow$ point. Now by composing these last two P.L. collapsings, and noting that $\partial d = d^r$, we get the desired P.L. collapsing $Y \cap \partial d \searrow$ point.

This completes Case 3.

CASE 4. Suppose $f_1 \in J'$ and $f_1 \cap g \neq f_1$. If in addition we have $J - J' = \emptyset$ then this case is deduced from Case 3 just as Case 2 (for $J - J' = \emptyset$) was deduced from Case 1. If $J - J' \neq \emptyset$ then this special case is dealt with exactly as in Case 2 (when $J - J' \neq \emptyset$).

This completes Case 4.

We are still not finished with the verification of the part of 7.6($k, \ell + 1$)(c) referred to in 7.2(e). For we still must show that any intersection

$$Y' = \left(\bigcap_{f \in J'} \partial Y(f) \right) \bigcap \left(\bigcap_{f \in J - J'} Y(f) \right)$$

is a P.L. cell or the empty set, where $J \subset S_{k,\ell+1}$ and $J' \subset J$, $J' \neq J$. If $Y' \neq \emptyset$ then we must have that $J \cap (S_{k,\ell+1} - S_{k,\ell}) = J''$ is either the empty set or contains only one simplex $f_1 \in J''$. If $J'' = \emptyset$ then 7.6(k, ℓ)(c) applies to show that Y' is a P.L. cell or the empty set. If $J'' = \{f_1\}$ then there are two cases ($f_1 \in J - J'$ or $f_1 \in J'$) to be considered. If $f_1 \in J - J'$ then set

$$X' = \left(\bigcap_{f \in J'} \partial Y(f) \right) \bigcap \left(\bigcap_{f \in J - J' - \{f_1\}} Y(f) \right) \cap \hat{f}_1.$$

Note that Y' is equal the total space of a bundle $\gamma(f_1)$ over X' having a P.L. cell for fiber. Since X' is a P.L. cell or the empty set (see 7.7(b), 7.6(k,ℓ)) it follows that Y' is a P.L. cell or the empty set. The verification that Y' is a P.L. cell in the case when $f_1 \in J'$ is left as an exercise.

This completes the verification of the induction step 7.6$(k,\ell) \Rightarrow$ 7.6$(k, \ell+1)$.

Note that 7.6$(n,m) \Rightarrow$ 7.2(a)(c)(d)(e). It is left as an exercise to show that the preceding induction argument (in Steps I, II above) can be carried out with sufficient metric control so that 7.2(b) will also be satisfied. [This entails for example choosing the $\delta, \varepsilon, \varepsilon', \varepsilon''$ of 7.8, 7.9, 7.10, 7.13, 7.14, 7.15, 7.16 so as to depend only on $(x, n, N(c), \mu_{n+m})$ of 7.2(b).]

This completes the proof of 7.2 when 7.4 is satisfied.

To prove 7.2 when 7.4 is not satisfied we proceed by induction as before. We note at each step 7.6$(k,\ell) \Rightarrow$ 7.6$(k,\ell+1)$ of the induction argument that metric control (as in 7.2(b)) may be assumed to hold. For this reason any linear construction in Step I, II of the induction step 7.6$(k,\ell) \Rightarrow$ 7.6$(k,\ell+1)$ has an "almost linear" analogue.

This completes the proof of lemma 7.2.

§8. TRIANGULATING IMAGE BALLS

Roughly speaking the main result of this section (see 8.5) tells us that the balls constructed in section 7 can be isotopied to new balls so that the images of these balls under the map $F^q : M \to M$ are triangulated by the triangulations referred to in proposition 6.8, for sufficiently large positive integers q.

We first give a construction of a finite collection of balls $\{B_{s,i} : i \in I_e\}$ in $X_{s,e}$ for each $e \in C$. These collections are referred to in lemma 8.2 and proposition 8.5 below. (See 6.7 for $X_{s,e}, X_{u,e}$.)

Define a cell structure E for R^m as follows. Set $D = \{(x_1, x_2, \ldots, x_m) \in R^m : 0 \leq x_i \leq 1\}$, and let $\mathbb{Z}^m$ denote all the points in R^m which have integer valued coordinates. Then E is the cell structure which has for its m-cells all the translates of D by vectors in $\mathbb{Z}^m$, and has for cells of dimension less than m all the translates of the faces of D by vectors in $\mathbb{Z}^m$. For any positive number σ set E_σ equal to the image of E under scalar multiplication by σ. Let $\{B_{\sigma,i} : i \in I\}$ denote the collection of all balls in R^m which have radius σ and are centered at the vertices of E_σ. For each $X_{s,e}$ let $\{B_{s,i} : i \in I_e\}$ denote the balls of $\{B_{\sigma,i} : i \in I\}$ which are contained in $X_{s,e}$ and are within a distance 3σ of $N_{s,e}$. For $t > 0$ and $i \in I_e$ let ${}_tB_{s,i}$ be the ball of radius σt having the same center as does $B_{s,i}$.

For future reference we list some of the properties that the collections $\{B_{s,i} : i \in I_e\}$ satisfy.

8.1. (a) The radius of each of the balls $\{B_{s,i} : i \in I_e, e \in C\}$ is the same number σ, which may be assumed as small as need be.

(b) Let $X \subset R^m$ be a subset of diameter $d\sigma$. For any given $e \in C$ the number of $\{{}_2B_{s,i} : i \in I_e\}$ which intersect X is less than $2^m(d+4)^m$.

(c) Given any compact subsets $\{N_{s,e} \subset X_{s,e} : e \in C\}$ it may be assumed that $N_{s,e} \subset \cup_{i \in I_e} \mathrm{Int}(B_{s,i})$ holds for all $e \in C$.

(d) The Lebesque covering number for the covering $\{B_{s,i} : i \in I_e\}$ of $N_{s,e}$ is denoted by σ'. The ratio σ'/σ depends only on m.

LEMMA 8.2. *There is an integer $\beta > 0$ that depends only on n, m and η of 6.1(e). Suppose in 7.2 that $x = \beta y$ for some integer $y > 0$. Let $0 < a_1 < a_2 < a_3 < \cdots < a_{n+1} < 2$ be a given increasing sequence of numbers, and let ${}_1N_{s,e} \subset {}_2N_{s,e} \subset \cdots \subset {}_{n+1}N_{s,e} \subset X_{s,e}$ be an increasing sequence of sets. Then for each $e \in C$ and each $k \in I_e$ there is an almost linear redundant ball structure $\{{}_kY_j(f) : f \in K_{u,e}, 1 \leq j \leq y\}$ for each $K_{u,e}$ which satisfy the following properties.*

(a) Each ${}_kY_j(f)$ *is one of the* $\{Y_i(f) : 1 \le i \le x\}$ *of 7.2.*

(b) If ${}_2B_{s,k}, {}_2B_{s,k'}$ *both intersect a set* Z *having diameter* $\le 4\eta\sigma$*, but* $k \ne k'$ *(for* $k' \in I_e$*), then the collections* $\{{}_kY_i(f) : f \in K_{u,e}, 1 \le i \le y\}$ *and* $\{{}_{k'}Y_i(f) : f \in K_{u,e}, 1 \le i \le y\}$ *have no members in common.*

(c) Suppose that for some $e_1, e_2 \in C$*,* $f \in K_{u,e}, k_1 \in I_{e_1}, k_2 \in I_{e_2}, k \in I_e, k' \in I_e, j$ *and* j' *in* $\{1, 2, \ldots, y\}$*, and some* a_i *we have that*

$$F^q(g_e({}_kY_j(f) \times B_{s,k})) \subset g_{e_1}({}_{n+1}N_{u,e_1} \times {}_{a_{n+1}}B_{s,k_1}),$$
$$F^q(g_e({}_{k'}Y_{j'}(f) \times B_{s,k'})) \subset g_{e_1}({}_{n+1}N_{u,e_1} \times {}_{a_{n+1}}B_{s,k_1}),$$
$$F^q(g_e({}_kY_j(f) \times B_{s,k})) \subset g_{e_2}({}_iN_{u,e_2} \times {}_{a_i}B_{s,k_2}),$$
$$F^q(g_e({}_{k'}Y_{j'}(f) \times B_{s,k'})) \not\subset g_{e_2}({}_iN_{u,e_2} \times {}_{a_i}B_{s,k_2}).$$

Then we must have that ${}_kY_j(f) \ne {}_{k'}Y_{j'}(f)$*.*

PROOF OF LEMMA 8.2: Note it follows from 8.1(b) that for each $e \in C$ there are subsets $\{J_{j,e} : 1 \le j \le (10\eta)^{2m}\}$ of the index set I_e in 8.1 which satisfy the following properties.

8.3. (a) The subsets $\{J_{j,e} : 1 \le j \le (10\eta)^{2m}\}$ are pairwise disjoint and their union is all of I_e.

(b) For any $j \in \{1, 2, \ldots, (10\eta)^{2m}\}$ and any $i, i' \in J_{j,e}$ we have that ${}_2B_{s,i}, {}_2B_{s,i'}$, are distance greater than $4\eta\sigma$ apart if $i \ne i'$.

For any $Y_i(f) \times B_{s,k}$, with $i \in \{1, 2, \ldots, x\}$, $k \in I_e$, $f \in K_{u,e}$, $e \in C$, we define the **pattern** of $Y_i(f) \times B_{s,k}$—denoted by $P(f, k, i)$—to be the collection of all the sets in $\{{}_{a_i}B_{s,k'} : k' \in I_{e'}, e' \in C, a_i \in \{a_1, a_2, \ldots, a_{n+1}\}\}$ which satisfy $F^q \circ g_e(Y_i(f) \times B_{s,k}) \subset g_{e'}({}_iN_{u,e'} \times {}_{a_i}B_{s,k'})$. We let $\mathcal{P}$ denote the collection of all such patterns. Note that it follows from 6.1(e), 6.7(b), and 8.1(b) that there are integers $\alpha_1 > 0$, $\alpha_2 > 0$ and subsets $\{\mathcal{P}_i : 1 \le i \le \alpha_2\}$ of $\mathcal{P}$ satisfying the following properties.

8.3. (c) α_1, α_2 depend only on $N(C)$, n, m, η.

(d) The subsets $\{\mathcal{P}_i : 1 \le i \le \alpha_2\}$ are pairwise disjoint, and $\cup_{i=1}^{\alpha_2} \mathcal{P}_i = \mathcal{P}$. Moreover if for some i we have $P_1, P_2 \in \mathcal{P}_i$ then either $P_1 = P_2$ or $P_1 \cap P_2 = \emptyset$.

(e) For any $e \in C$, $f \in K_{u,e}$, $k \in I_e$ there are at most α_1 different patterns in the collection $\{P(f, k, i) : 1 \le i \le x\}$.

Now we set

8.3. (f) $\beta = (10\eta)^{2m}\alpha_1\alpha_2$.

Note it follows from 8.3(c), 6.2(c) that β depends only on n, m, η as is required in 8.2. Note also that it follows from 8.3(e)(f) and the equality $x = \beta y$ that for each $e \in C$, $f \in K_{u,e}$, $k \in I_e$ we can choose a pattern in $\{P(f, k, i) : 1 \le i \le x\}$—denoted by $P(f, k)$—so that the following property holds.

8.3. (g) After reordering the $\{Y_i(f) : 1 \le i \le x\}$ (if need be) we will have that

$$P(f,k,i) = P(f,k)$$

for all $i \in \{1,2,\ldots,(10\eta)^{2m}\alpha_2 y\}$.

Now choose subsets $\{V_i : 1 \le i \le (10\eta)^{2m}\alpha_2\}$ of the union $\cup_{e\in C} K_{u,e} \times I_e$ to satisfy the following.

8.3. (h) The $\{V_i : 1 \le i \le (10\eta)^{2m}\alpha_2\}$ are pairwise disjoint, and

$$\bigcup_{i=1}^{(10\eta)^{2m}\alpha_2} V_i = \bigcup_{e\in C} K_{u,e} \times I_e.$$

(i) Suppose (f,k), $(f',k') \in V_i$ for some i, and $f \in K_{u,e}$, $f' \in K_{u,e'}$. Then there is a $j \in \{1,2,\ldots,\alpha_2\}$ such that both of $P(f,k)$, $P(f',k')$ lie in $\mathcal{P}_j$. Moreover there is $t \in \{1,2,\ldots,(10\eta)^{2m}\}$ such that $k \in J_{t,e}$ and $k' \in J_{t,e'}$.

To complete the proof of 8.2 we choose the $\{{}_kY_i(f) : f \in K_{u,e},\ e \in C,\ k \in I_e, \text{ and } 1 \le i \le y\}$ as follows. If $(f,k) \in V_q$ then set

$${}_kY_i(f) = Y_{y(q-1)+i}(f).$$

Clearly 8.2(a) is satisfied. Properties 8.2(b)(c) follow from 8.3(a)(b)(d)(g)(h)(i).

This completes the proof of lemma 8.2.

We can now state the main result of this section. We let $q, C, K_{u,e}$, $Y_i(f), \{B_{s,j} : j \in I_e\}, N_{s,e}, \{{}_kY_i(f) : 1 \le i \le y\}$ be as in 6.1, 6.2, 7.2, 8.1, 8.2. For each $e \in C$ and each $j \in \{1,2,\ldots,n+1\}$ we choose a compact subset ${}_jN_{u,e}$ of $N_{u,e}$ (see 6.8(a) for $N_{u,e}$) so that the following properties hold. (Here n, m are as in 6.1.)

8.4. ${}_jN_{u,e} \subset \mathrm{Int}({}_{j+1}N_{u,e})$ for $j \in \{1,2,\ldots,n\}$; ${}_{n+1}N_{u,e} = N_{u,e}$;

$$e \subset \mathrm{Int}\left(\cup_{e'\in e}\, g_{e'}({}_1N_{u,e'} \times X_{s,e'})\right).$$

In the rest of this chapter we assume that the $\{{}_iN_{u,e}\}$ of 8.2 and of 8.4 are identical.

For any $e, e' \in C$, $k' \in I_{e'}$, $i \in \{0,1,2,\ldots,y\}$, $f \in K_{u,e'}$, set ${}_{k'}Y_{i,e}(f) = \emptyset$ if neither $e \subset e'$ nor $e' \subset e$ hold. Otherwise let ${}_{k'}Y_{i,e}(f)$ denote the image of ${}_{k'}Y_i(f)$ under the composite map

$$R^n = R^n \times 0 \xrightarrow{g_{e'}} M \xrightarrow{g_e^{-1}} R^n \times R^m \xrightarrow{\text{proj.}} R^n.$$

This same recipe defines ${}_{k'}Y'_{i,e}(f)$ in 8.5(b) below.

PROPOSITION 8.5. *If ε, σ of 6.8, 8.1 are chosen sufficiently small, and if q and the r_i (of 6.8) are chosen sufficiently large, then for each $e \in C$, $f \in K_{u,e}$, $i \in \{1,2,\ldots,y\}$, and each $k \in I_e$, there is a subset ${}_kY_i'(f) \subset X_{u,e}$ satisfying all the following properties.*

(a) Let $k' \in I_e$, $f' \in K_{u,e}$, and $i' \in \{1,2,\ldots,y\}$. Then ${}_kY_i'(f) \cap f' \neq \emptyset$ if and only if ${}_kY_i(f) \cap f' \neq \emptyset$. Also ${}_kY_i'(f) \cap {}_{k'}Y_{i'}'(f') \neq \emptyset$ if and only if ${}_kY_i(f) \cap {}_{k'}Y_{i'}(f') \neq \emptyset$.

(b) For each $e \in C$, $k \in I_e$ there is a homeomorphism $h_k : R^n \to R^n$. If for some $e' \in C$, $f \in K_{u,e'}$, $i \in \{0,1,2,\ldots,y\}$, and $k' \in I_{e'}$ we have

$$g_{e'}({}_{k'}Y_i(f) \times {}_2B_{s,k'}) \cap g_e(X_{u,e} \times {}_2B_{s,k}) \neq \emptyset,$$

then we must also have that

$$h_k({}_{k'}Y_{i,e}(f)) = {}_{k'}Y_{i,e}'(f).$$

(c) Given $e, e' \in C$, $k \in I_e$, $i \in \{1,2,\ldots,y\}$, and $f \in K_{u,e}$ suppose that $F^q(g_e({}_kY_i'(f) \times B_{s,k})) \subset g_{e'}({}_1N_{u,e'} \times N_{s,e'})$. Then $\rho_1 \circ g_{e'}^{-1} \circ F^q \circ g_e({}_kY_i'(f) \times B_{s,k})$ is a subcomplex of $K_{u,e'}$, where $\rho_1 : R^n \times R^m \to R^n$ is projection onto the first factor.

(d) If $F^q \circ g_e({}_kY_i'(f) \times B_{s,k}) \not\subset \cup_{e' \in C}\, g_{e'}(X_{u,e'} \times X_{s,e'})$, then we have ${}_kY_i'(f) = {}_kY_i(f)$.

(e) Let β be the least upper bound for the diameters of all the balls $\{{}_kY_i(f)\}$ of 8.2. Then $|h_k(x) - x| \leq \phi\eta^2\beta a^{-1}\lambda^{-q}$, where a, λ, η come from 5.1, 6.1, and where $\phi > 1$ is independent of β, q.

PROOF OF PROPOSITION 8.5: We choose the $\{N_{s,e} : e \in C\}$ from 8.1(c) so that $e \subset \text{Int}\,(\cup_{e' \in e}\, g_{e'}({}_1N_{u,e'} \times N_{s,e}))$ (compare with 8.4).

Let S denote the collection of all $\{B_{s,i} : e \in C,\ i \in I_e\}$.

Note it follows from 8.1, 6.1(e), 6.2(c) that there is an integer $\alpha > 0$, and subsets $\{S_{i,j} : 0 \leq i \leq n+m,\ 1 \leq j \leq \alpha\}$ of S which satisfy the following.

8.6 (a) α depends only on n, m, and η of 6.1(e).

(b) If $B_{s,k} \in S_{i,j}$ and $k \in I_e$ then $\dim(e) = i$. Moreover we have:

$$S = \cup_{i,j}\, S_{i,j};$$
$$S_{i,j} \cap S_{i',j'} = \emptyset \text{ if } i \neq i' \text{ or if } j \neq j'.$$

(c) If for some i, j we have that the $B_{s,k}$ and $B_{s,k'}$ are both in $S_{i,j}$ (for $k \in I_e$ and $k' \in I_{e'}$) then there is no third $B_{s,k''} \in S$ (with $k'' \in I_{e''}$) such that both

$$g_e(X_{u,e} \times {}_2B_{s,k}) \cap g_{e''}(X_{u,e''} \times {}_2B_{s,k''}) \neq \emptyset,$$

and

$$g_{e'}(X_{u,e'} \times {}_2B_{s,k'}) \cap g_{e''}(X_{u,e''} \times {}_2B_{s,k''}) \neq \emptyset,$$

are satisfied.

Set $S^{i,j} \equiv \cup S_{a,b}$, where the union runs over all (a,b) with either $a < i$, or $a = i$ and $b \leq j$. Our proof of 8.5 is an induction argument over the increasing sequence $\ldots, S^{i,j}, S^{i,j+1}, \ldots$. Before stating the induction hypothesis we need some more notation. For each $\ell \in \{0,1,2,\ldots,n+m\}$ and each $p \in \{1,2,\ldots,\alpha\}$ let $A_{\ell,p}$ denote the set of all ${}_kY_i(f)$ such that for some $B_{s,k'} \in S^{\ell,p}$ we have that $F^q \circ g_e({}_kY_i(f) \times B_{s,k}) \subset g_{e'}({}_1N_{u,e'} \times B_{s,k'})$, where $f \in K_{u,e}$, $k \in I_e$, and $k' \in I_{e'}$. Let $A^-_{\ell,p}$ denote all the ${}_kY_i(f)$ which are not in $A_{\ell,p}$. Set $A_{0,0} = \emptyset$. For each $B_{s,j} \in S - S^{\ell,p}$ (with $j \in I_e$) and each $t \in \{0,1,\ldots,n+1\}$ define a subset $A^-_{\ell,p,j,t}$ of $A^-_{\ell,p}$ as follows. Set $A^-_{\ell,p,j,0} = \emptyset$. For ${}_kY_i(f) \in A^-_{\ell,p}$—with $f \in K_{u,e'}$—we have ${}_kY_i(f) \in A^-_{\ell,p,j,t}$ if and only if ${}_kY_i(f) \notin \cup_{i=1}^{t-1} A^-_{\ell,p,j,i}$ and $F^q \circ g_{e'}({}_kY_i(f) \times B_{s,k}) \subset g_e({}_tN_{u,e} \times {}_{a_t}B_{s,j})$. (Here σ is as in 8.1 and for any $t \in \{1,2,\ldots,n+1\}$ we set $a_t = (1+(t-1)/2n)\sigma + \sigma$.) In the rest of this proof we assume that the $a_1, a_2, \ldots, a_{n+1}$ just defined are the same as the $a_1, a_2, \ldots, a_{n+1}$ of 8.2.

We can now state the induction hypothesis.

8.7(ℓ,p). Each of the ${}_kY_i(f)$ has been isotopied to a new position in R^n—denoted by ${}_kY_i'(f)$—such that the following hold.

(a) The ${}_kY_i'(f)$ satisfy 8.5(a)(b)(d).

(b) All the ${}_kY_i'(f)$ for which ${}_kY_i(f) \in A_{\ell,p}$ satisfy 8.5(c).

(c) There is a positive integer $r_{\ell,p}$ which depends only on the $(n+m, \ell, p, N(C), \eta, x, r_{\ell-1}, \mu_{\ell-1})$ and satisfies $r_{\ell-1} < r_{\ell,p} < r_\ell$. (Here x comes from 8.2, (r_i, μ_i) come from 6.8, and η comes from 6.1(e).) If for some $B_{s,k'} \in S^{\ell,p}$, with $k' \in I_{e'}$ and $\dim(e') = \ell$, and for some ${}_kY_i(f) \in A_{\ell,p}$ we have that $F^q \circ g_e({}_kY_i'(f) \times B_{s,k}) \subset g_{e'}({}_1N_{u,e'} \times B_{s,k'})$—where $f \in K_{u,e}$—then $\rho_1 \circ g_{e'}^{-1} \circ F^q \circ g_e({}_kY_i'(f) \times B_{s,k})$ must be a subcomplex of $L_{u,e'}^{(r_{\ell,p})}$. Here $\rho_1 : R^n \times R^m \to R^n$ is the projection onto the first factor, and $L_{u,e'}^{(r_{\ell,p})}$ is the almost linear $r_{\ell,p}$-derived subdivision of $L_{u,e'}$ which is subdivided by the almost linear r_ℓ-derived subdivision $L_{u,e'}^{(r_\ell)}$ of $L_{u,e'}$ given in 6.8(d).

(d) Each ${}_kY_i'(f)$ is an almost linear polyhedron having just one n-dimensional member. For any $A^-_{\ell,p,j,t}$ (with $B_{s,j} \subset S - S^{\ell,p}$ and $j \in I_e$) denote by $A'_{\ell,p,j,t}$ the collection of almost linear polyhedra $\{\rho_1 \circ g_e^{-1} \circ F^q \circ g_{e'}({}_kY_i'(f) \times B_{s,k}) : {}_kY_i(f) \in A^-_{\ell,p,j,t}, f \in K_{u,e'}\}$. For any subset $J \subset \cup_{t=1}^{n+1} A'_{\ell,p,j,t}$ we require that all the almost linear polyhedra in J are in transverse position to one another and generate an almost linear polyhedron denoted by V_J. We also require that each V_J is in transverse position to $K_{u,e}$ and that V_J, $K_{u,e}$ generate an almost linear polyhedron denoted by W_J.

(e) The following metric conditions must hold for all the $e, t, B_{s,j}, W_J$, $V_J, L_{u,e}$ of (d). There is $\delta_{\ell,p} > 0$, which depends only on the $(\ell, p, q, n+m, N(C), x, a, \lambda, \eta, \mu)$ such that $\delta_{\ell,p}$ is a lower bound for all the $\tau(X)$, $D(X)/D(Y)$ where X, Y, can be any of the $V_J, W_J, L_{u,e}$. (The numbers a, λ, μ come from 5.1(b) and 7.2(b).) There is a number $\alpha_{\ell,p} > 0$ such that $\alpha_{\ell,p} >> \overline{D}(L_{u,e})$ and such that the collection of all distinct polyhedra in $\cup_{t=1}^{n+1} A'_{\ell,p,j,t}$ are in $\alpha_{\ell,p}$-transverse position to one another. Finally there is $\gamma_{\ell,p} > 0$, which depends only on the $(\ell, p, N(C), n+m, \eta, r_{\ell-1}, \mu_{\ell-1})$ such that $V_J \# \overline{D}(L_{u,e}) \leq \gamma_{\ell,p}$ for each V_J.

This completes the statement of the induction hypothesis in our argument. Before completing the induction argument we need to discuss how the $r_0, \ldots, r_{n+m}, q, \sigma, \varepsilon$, which are used in 8.5, are selected.

In order of dependence we first choose the $r_0, \ldots, r_{n+m}$, secondly choose q, thirdly choose σ, and lastly choose ε so that the following hold.

8.8 (a) We must choose each r_ℓ sufficiently large so that for all $p \in \{1, 2, \ldots, \alpha\}$, we have that $r_{\ell,p} < r_\ell$ in $8.7(\ell, p)$. (This is possible since $r_{\ell,p}$ is independent of our choice for r_ℓ by $8.7(\ell, p)$(c)).

(b) We must choose q sufficiently large so that $a\lambda^{-q} << \mu\eta^{-2}$ and $a\lambda^{-q} << (\sigma'/\sigma)\eta^{-2}$. (Note the q depends on the $r_0, r_1, \ldots, r_{n+m}$, because μ of 7.2(b) depends on the $r_0, r_1, \ldots, r_{n+m}$, however q is independent of σ, σ' because σ'/σ of 8.1(d) depends only on m.)

(c) Let $z > 0$ denote the minimum over all $e, e' \in C$, with neither of $e \subset e'$, $e' \subset e$ true, of the distance from $g_e(N_{u,e} \times N_{s,e})$ to $g_{e'}(N_{u,e'} \times N_{s,e'})$. Choose $\sigma << z\eta^{-2}a^{-1}\lambda^{-q}$.

(d) Let $z' > 0$ denote the minimum over all $e \in C$ and all $j \in \{1, 2, 3, \ldots, n\}$ of the distance from ${}_jN_{u,e}$ to the boundary of ${}_{j+1}N_{u,e}$ or of the distance from $g_e(e)$ to the boundary of $\cup_{e' \in e} g_{e'}({}_1N_{u,e'} \times N_{s,e'})$. Choose $\varepsilon << a^{-1}\lambda^{-q}\eta^{-2}z'$.

REMARK. The most important outcome achieved by choosing q sufficiently large in 8.8(b), and by choosing σ sufficiently small in 8.8(c), is to assure the existence of numbers $\alpha_{0,0}, \gamma_{0,0}$ satisfying 8.7(0,0)(e).

In order to get the induction argument under way we must first verify 8.7(0,0). Note if we set ${}_kY_i'(f) = {}_kY_i(f)$ for all i, k, f then the ${}_kY_i'(f)$ will satisfy all of 8.7(0,0)(a)(b)(c). These ${}_kY_i'(f)$ will also satisfy 8.7(0,0)(d)(e) with the following exceptions: in 8.7(0,0)(d) some V_J may not be in transverse position to $K_{u,e}$ (compare with 7.2(a),

8.2(a), 8.8(c)); in 8.7(0,0)(e) X, Y are not allowed to be W_J (compare with 7.2(a)(b), 8.8(c), 8.2(a)). To correct this deficiency of the present $\{{}_kY_i'(f)\}$ we isotopy each ${}_kY_i'(f)$ to a new position—also denoted by ${}_kY_i'(f)$—so that 8.7(0,0)(a)(b)(c) remain satisfied for the $\{{}_kY_i'(f)\}$ and so that 8.7(0,0)(d)(e) become satisfied.

Towards this end we triangulate each ${}_kY_i(f)$ by an almost linear triangulation ${}_kT_i(f)$ by using 7.5 and 8.2(a). (By 7.5 and 8.2(a) each ${}_kY_i(f) = \Delta_1 \times \Delta_2$, where Δ_1, Δ_2 are simplices; let ${}_kT_i(f)$ be the barycentric subdivision of the product cell structure $\Delta_1 \times \Delta_2$.) Each of the desired new ${}_kY_i'(f)$ is going to be the end result of applying a P.S. flow to the triangulation ${}_kT_i(f)$. (More precisely let $\partial_kT_i(f)$ denote the subcomplex of ${}_kT_i(f)$ satisfying $|\partial_kT_i(f)| = \partial|{}_kT_i(f)|$ and let $\partial_kT_i'(f)$ denote the triangulation to which $\partial_kT_i(f)$ is moved under the flow. Then the polyhedron ${}_kY_i'(f)$ consists of all the triangles in $\partial_kT_i'(f)$ and the compact n-dimensional set bounded by $|\partial_kT_i'(f)|$.) These flows, and thus the new ${}_kY_i'(f)$, are constructed by induction over the sequence $\ldots,(\ell,p),(\ell,p+1),\ldots,(\ell,\alpha),(\ell+1,1),(\ell+1,2),\ldots$. Suppose that the flows have been constructed so that for any $B_{s,j} \in S^{\ell,p}$ we have that 8.7(d) holds for the $\{A^-_{0,0,j,t} : 1 \le t \le n+1\}$. Now (using almost linear versions of lemma 3.1) we can, by further flowing only those ${}_kY_i'(f)$ with ${}_kY_i(f) \in A^-_{0,0,k,t}$ for some $B_{s,k} \in S_{\ell,p+1}$ and some $t \in \{1,2,\ldots,n+1\}$, obtain a new collection of $\{{}_kY_i'(f)\}$ which satisfy 8.7(0,0)(d) for all $A^-_{0,0,k,t}$ with $B_{s,k} \in S^{\ell,p+1}$.

Note that if the flows of the last paragraph are sufficiently close to the identity then the almost linear version of lemma 3.7 (as strengthened in Remark 3.7′) may be applied several times at each step of the induction argument of the preceding paragraph to show that the new $\{{}_kY_i'(f)\}$ still satisfy 8.7(0,0)(a). Since 8.7(0,0)(b)(c) are vacuous statements they also still remain satisfied.

Note also that the preceding P.S. flow for each ${}_kT_i(f)$—denoted by $({}_kT_i(f))_t$, $t \in [0,1]$—can be selected so as to satisfy the following.

<u>8.9.</u> If ${}_kY_i(f)$ and ${}_{k'}Y_i'(f)$ are always in the same sets $A^-_{0,0,j,t}$ of 8.7(0,0)(d) (for all j, t) and ${}_kY_i(f) = {}_{k'}Y_{i'}(f)$ then we have

$$({}_kT_i(f))_t = ({}_{k'}T_{i'}(f))_t$$

for all $t \in [0,1]$. (Compare with 8.2(a), 8.8(c).)

It is left as an exercise to show that the P.S. flows $\{({}_kT_i(f))_t,\ t \in [0,1]\}$ can be chosen so that 8.7(0,0)(e) is also satisfied. Note that 8.9 and 8.2(c) are of essential importance in verifying this last claim.

In order to complete our induction proof of 8.4 we must carry out the induction step by showing $8.7(\ell,p) \Rightarrow 8.7(\ell,p+1)$. (Of course if $p=\alpha$ then the induction step consists of showing $8.7(\ell,p) \Rightarrow 8.7(\ell+1,1)$.

Choose any $B_{s,j} \in S_{\ell,p+1}$ (with $j \in I_e$). For this j,e and any $t \in \{1,2,\dots,n+1\}$ let K_{t-1} denote the almost linear polyhedron generated by all the almost linear polyhedra in $A'_{\ell,p,j,t}$, and let $\{K_{t-1,k} : k \in J_{t-1}\}$ denote all the polyhedra in $A'_{\ell,p,j,t}$. Set $K = K_0$. For each $B_{s,j'} \in S - S^{\ell,p+1}$, with $j' \in I_{e'}$, and for each $1 \leq t \leq n+1$, define a collection of polyhedra $A'_{\ell,p+1,j',t,e}$ as follows. If $g_e(X_{u,e} \times {}_2B_{s,j}) \cap g_{e'}(X_{u,e'} \times {}_2B_{s,j'}) = \emptyset$ then $A'_{\ell,p+1,j',t,e} = \emptyset$. Otherwise $A'_{\ell,p+1,j',t,e}$ consists of $K_{u,e,e'}$ and all the polyhedra $\{\rho_1 \circ g_e^{-1} \circ g_{e'}(Q \times B_{s,j'})\}$, where $Q \in A'_{\ell,p+1,j',t}$ but $\rho_1 \circ g_e^{-1} \circ g_{e'}(Q \times B_{s,j}) \notin \cup_{i=1}^{n} A'_{\ell,p,j,i}$, and where $\rho_1 : R^n \times R^m \to R^n$ is projection onto the first factor. Let $P_{e,j'}$ denote the polyhedron generated by the $\cup_{t=1}^{n+1} A'_{\ell,p+1,j',t,e}$.

Let $\{K_t : n+1 \leq t \leq b\}$ denote all the $\{P_{e,j'}\}$. Set $L = L_{u,e}^{(r_{\ell,p})}$.

We now apply the almost linear version of proposition 2.14 (see 4.9) to the $K, L, \{K_t : 1 \leq t \leq b\}$, $\{K_{t,k} : 1 \leq t \leq n,\ k \in J_t\}$ of the preceding paragraph to get homeomorphisms $f_t : R^n \to R^n$, $t = 1,2,\dots,n+1$, satisfying the almost linear versions of 2.14(a)-(d). Let $L_{u,e}^{(r_{\ell,p+1})}$ be the β-fold derived subdivision of L which is subdivided by $L_{u,e}^{(r_\ell)}$ of 6.8(d), where β comes from 2.14. [Thus by 2.14(b) we have that for each $e \in K$ the set $f_1(e)$ is a subcomplex of $L_{u,e}^{(r_{\ell,p+1})}$.] We perform this construction for each $B_{s,j} \in S_{\ell,p+1}$

Now we can define the new $\{{}_kY_i'(f)\}$ which satisfy $8.7(\ell,p+1)$ as follows. If ${}_kY_i(f) \in A^-_{\ell,p,j,t}$ for some $B_{s,j} \in S_{\ell,p+1}$ with $j \in I_e$ as in the preceding two paragraphs then set the new ${}_kY_i'(f)$—i.e., the ${}_kY_i'(f)$ of $8.7(\ell,p+1)$—equal to the image of the old ${}_kY_i'(f)$ under the homeomorphism $f_t : R^n \to R^n$ of the preceding paragraph. (More precisely we set the new $|{}_kY_i'(f)|$ equal to the image under f_t of the old $|{}_kY_i'(f)|$. Now we take for the partition—sets in the new ${}_kY_i'(f)$ all the sets of K_t' (see 2.14(c) for K_t') which are contained in $f_t(\partial|{}_kY_i'(f)|)$, together with the set $f_t(|{}_kY_i'(f)|$.) On the other hand if ${}_kY_i(f)$ is not contained in any $A^-_{\ell,p,j,t}$ as above, then set the new ${}_kY_i'(f)$ equal to the old ${}_kY_i'(f)$.

It is left as an exercise to verify that the new $\{{}_kY_i'(f)\}$ now satisfy $8.7(\ell,p+1)$. (Compare with 8.6, 2.14(a)-(d), $8.7(\ell,p)$, and 8.8(b)(d).)

Let $\{{}_kY_i'(f)\}$ denote the sets of $8.7(n,\alpha)$. These $\{{}_kY_i'(f)\}$ will satisfy all the conclusions of 8.5. (Compare with 8.8(b).)

This completes the proof of proposition 8.5.

§9. THE THICKENING THEOREM

In this section we construct for each of the ${}_kY_j(f)$ in 8.2, and for each positive integer t, a t-fold thickening of ${}_kY_j(f)$—denoted by ${}_kY_j(f;t)$—such that ${}_kY_j(f;t) \subset {}_kY_j(f;t+1)$ for all t. The thickening theorem (see 9.6) asserts roughly that for any t with $1 \leq t < \infty$ the collection of all $\{{}_kY_j(f;t) \times B_{s,k}\}$ is homeomorphic to the collection of all $\{{}_kY_j(f) \times B_{s,k}\}$.

Before beginning the construction of the thickened balls we must first introduce some more subsets of the $X_{u,e}, X_{s,e}$. For each $e \in C$ we let ${}_0N_{u,e} \subset X_{u,e}$ and ${}_0N_{s,e} \subset X_{s,e}$ be given subsets. There is no loss of generality in assuming that these subsets satisfy the following properties.

9.0. (a) ${}_0N_{u,e}, {}_0N_{s,e}$ are arbitrarily large compact subsets of $X_{u,e}$, $X_{s,e}$.

(b) ${}_0N_{u,e} \subset \mathrm{Int}({}_1N_{u,e})$ and ${}_0N_{s,e} \subset \mathrm{Int}(N_{s,e})$ hold for all $e \in C$. (Here $N_{s,e}$ comes from 8.1 and ${}_1N_{u,e}$ comes from 8.2, 8.4.)

Now for each ${}_kY_j(f)$, and each $K_{u,e'} \times B_{s,k'}$ with $k' \in I_{e'}$ (see 8.1 for $I_{e'}$) we select a subset $H(k,j,f;k')$ of the $\{{}_qY_i(f') : f' \in K_{u,e'}, q \in I_{e'}, 1 \leq i \leq y\}$ as follows.

9.1. (a) If $F^q \circ g_e({}_kY_j(f) \times {}_2B_{s,k})$ is not contained in the intersection $g_{e'}({}_0N_{u,e'} \times {}_0N_{s,e'}) \cap g_{e'}(X_{u,e'} \times {}_{3/2}B_{s,k'})$—where $k \in I_e$ and $f \in K_{u,e}$—then we set $H(k,j,f;k')$ equal the empty set.

(b) Otherwise note that the image of $\partial_k Y_i'(f) \times 0$ under the composite map

$$R^n \times 0 \xrightarrow{F^q \circ g_e} M \xrightarrow{g_{e'}^{-1}} R^n \times R^m \xrightarrow{\text{proj.}} R^n$$

is a subcomplex of $K_{u,e'}$ denoted by $K(k,j,f;k')$ (see 9.0, 8.5(c)). For each simplex $d \in K(k,j,f;k')$ we choose exactly one of the balls $\{{}_{k'}Y_i(d) : 1 \leq i \leq y\}$ and define $H(k,j,f;k')$ to be the collection of all these choices. Thus there is exactly one ball in $H(k,j,f;k')$ for each $d \in K(k,j,f;k')$.

LEMMA 9.2. *Suppose the y of 8.2 satisfies $y \geq (n+m+1)n(10\eta)^{2m}$. Then the $H(k,j,f;k')$ can be chosen to satisfy the following. If for any ${}_{k_1}Y_{j_1}(f_1)$ and ${}_{k_2}Y_{j_2}(f_2)$ we have that $g_{e_i}({}_{k_i}Y_{j_i}(f_i) \times {}_2B_{s,k_i}) \cap g_e({}_kY_j(f) \times {}_2B_{s,k}) \neq \emptyset$—for some ${}_kY_j(f)$, $k_i \in I_{e_i}$, $f_i \in K_{u,e_i}$, $i = 1,2$— then we* *must have that* $H(k_1, j_1, f_1; k') \cap H(k_2, j_2, f_2; k') = \emptyset$ *for all* k', *provided* $(f_1, k_1, j_1) \neq (f_2, k_2, j_2)$.

PROOF OF LEMMA 9.2: Set $S = \{{}_kY_i(f) : e \in C, k \in I_e, f \in K_{u,e}, 1 \leq i \leq y\}$. For each $j \in \{0, 1, \ldots, n+m\}$ let S_j denote the subset of all ${}_kY_i(f) \in S$ such that $k \in I_e$ with $\dim(e) = j$. Note that S is equal the union $\cup_{j=0}^{n+m} S_j$. It follows from 8.1(b) that each S_j is equal a union $\cup_{\ell=1}^{(10\eta)^{2m}} S_{j,\ell}$ of pairwise disjoint subsets $S_{j,\ell} \subset S_j$ which satisfy the following property.

9.3. Suppose ${}_{k_1}Y_{j_1}(f_1), {}_{k_2}Y_{j_2}(f_2) \in S_{j,\ell}$ for some j, ℓ. Then if $g_{e_i}({}_{k_i}Y_{j_i}(f_i) \times {}_2B_{s,k_i}) \cap g_e({}_kY_j(f) \times {}_2B_{s,k}) \neq \emptyset$—for $i = 1, 2$ and some ${}_kY_j(f) \in S$—then we must have $e_1 = e_2$ and $k_1 = k_2$ (where $k_1 \in I_{e_1}, k_2 \in I_{e_2}, k \in I_e$).

The first restriction we place on the selection of the sets $H(k, j, f; k')$ is the following.

9.4. (a) Suppose ${}_kY_j(f) \in S_{a,b}$. Then each ${}_{k'}Y_i(f')$ in $H(k, j, f; k')$ must satisfy

$$(b-1)n + an(10\eta)^{2m} < i \leq bn + an(10\eta)^{2m}.$$

The second restriction we place on the selection of the sets $H(k, j, f; k')$ is the following.

9.4. (b) Suppose ${}_{k_1}Y_{j_1}(f_1), {}_{k_2}Y_{j_2}(f_2) \in S_{a,b}$, $k_1 = k_2$ but $(f_1, k_1, j_1) \neq (f_2, k_2, j_2)$. Then if $g_{e_i}({}_{k_i}Y_{j_i}(f_i) \times {}_2B_{s,k_i}) \cap g_e({}_kY_j(f) \times {}_2B_{s,k}) \neq \emptyset$ (for $i = 1, 2$ and some ${}_kY_j(f) \in S$) we must also have $H(k_1, j_1, f_1; k') \cap H(k_2, j_2, f_2; k') = \emptyset$ for all k'.

Note that 9.3, 9.4 together imply the conclusion of lemma 9.2. So to complete the proof of 9.2 it remains to verify that 9.4 is possible. [Note that 9.4(a) is obviously possible.]

Towards verifying 9.4(b) we consider the collection of all $\{K(k, j, f; k') : {}_kY_j(f) \in S_{a,b}\}$ for fixed a, b, k', k. These are subcomplexes of $K_{u,e'}$—where $k' \in I_{e'}$—whose underlying sets $K(k, j, f; k')$ are P.L. $(n-1)$-manifolds. Moreover all the $\{K(k, j, f; k') :_k Y_j(f) \in S_{a,b}\}$ are in P.L. transverse position to one another (see 8.2(b), 8.5(b), 7.2(a)). It follows from these properties that for each simplex $d \in K_{u,e'}$ there are at most n of the subcomplexes $\{K(k, j, f; k') : {}_kY_j(f) \in S_{a,b}\}$ which contain d. Let $K\{k_i, j_i, f_i; k'), 1 \leq i \leq n$, denote these n subcomplexes. Now select the ball ${}_{k'}Y_{i'}(d)$ to belong to the collection $H(k_i, j_i, f_i; k')$, where $i' = (b-1)n + an(10\eta)^{2m} + i$.

This completes the proof of lemma 9.2.

9.5. THE THICKENINGS OF ${}_kY_j(f)$. For each ${}_kY_j(f)$, k' let $H(k,j,f;k')$ be as in 9.2. Set ${}_{k',k}Y_j(f;1)$ equal the subset of R^n which is mapped under the composite map of 9.1 onto the union of all ${}_{k'}Y_q'(f')$ such that ${}_{k'}Y_q(f') \in H(k,j,f;k')$. Define ${}_kY_j(f;1)$ to be the union of ${}_kY_j'(f)$ with all the ${}_{k',k}Y_j(f;1)$. We call ${}_kY_j(f;1)$ the **first level thickening** of ${}_kY_j(f)$.

The higher order thickenings of ${}_kY_j(f)$ are defined by induction over the positive integer t. Suppose the **tth level thickening** ${}_kY_j(f;t)$ has been defined for each ${}_kY_j(f)$. For each ${}_kY_j(f), k'$ let $H^t(k,j,f;k')$ denote the collection of all ${}_{k'}Y_i(d;t)$ such that ${}_{k'}Y_i(d) \in H(k,j,f;k')$. Set ${}_{k',k}Y(f;t+1)$ equal to the subset of R^n which is mapped under the composite map of 9.1 onto the union of all sets in $H^t(k,j,f;k')$. Define ${}_kY_j(f;t+1)$ equal to the union of ${}_kY_j'(f)$ with all the ${}_{k',k}Y_j(f;t+1)$.

Note that ${}_kY_j(f;t) \subset {}_kY_j(f;t+1)$ holds for all t,k,j,f.

Before stating the next theorem it is convenient to introduce some more notation. Let $e,e' \in C$, $k' \in I_{e'}$, $f' \in K_{u,e'}$, $i' \in \{1,2,\ldots,y\}$, and let t denote a positive integer. Define a subset ${}_{k'}Y_{i',e}(f';t)$ of R^n to be empty if neither $e' \subset e$ nor $e \subset e'$ holds. Otherwise define ${}_{k'}Y_{i',e}(f';t)$ to be the image of ${}_{k'}Y_{i'}(f';t)$ under the composite map

$$R^n = R^n \times 0 \xrightarrow{g_{e'}} M \xrightarrow{g_e^{-1}} R^n \times R^m \xrightarrow{\text{proj.}} R^n.$$

For any $e \in C$, $k \in I_e$ let $S_{k,e}$ denote all the ${}_{k'}Y_{i'}(f')$—with $k' \in I_{e'}$, $e' \in C$, $f' \in K_{u,e'}$—which satisfy $g_{e'}({}_{k'}Y_{i'}(f') \times {}_2B_{s,k'}) \cap g_e({}_0N_{u,e} \times {}_2B_{s,k}) \neq \emptyset$; and set $T_{k,e} = \cup_{k'} Y_{i',e}(f')$, where the union runs over all ${}_{k'}Y_{i'}(f') \in S_{k,e}$.

THICKENING THEOREM 9.6. *Suppose that for some $e \in C$, $k \in I_e$ we have that ${}_0N_{s,e} \cap {}_2B_{s,k} \neq \emptyset$. Then for any integer $t \geq 1$ there is an embedding $h_{k,t} : T_{k,e} \to R^n$ such that*

$$h_{k,t}({}_{k'}Y_{i',e}(f')) = {}_{k'}Y_{i',e}(f';t)$$

holds for all ${}_{k'}Y_{i'}(f') \in S_{k,e}$.

The proof of theorem 9.6 is given in section 11 below.

§10. RESULTS IN P.L. TOPOLOGY

In this section we review some results in P.L. topology, and prove two P.L. lemmas (see 10.7, 10.21) which are used in the proof of the thickening theorem in section 11. The reader is referred to [12] for a more detailed account of P.L. topological techniques.

We begin by recalling the existence and uniqueness theorem for regular neighborhoods. Let $(M, \partial M)$ denote a P.L. manifold pair and $X \subset M$ a compact P.L. subset of M. A **regular neighborhood** for X in M consists of another compact P.L. subset N in M satisfying

(a) N is a neighborhood for X in M,

(b) N collapses to X, i.e., $N \searrow X$ (see page 88 in [12]).

(c) N is a P.L. manifold.

The regular neighborhood N is said to meet ∂M **regularly** if $N \cap \partial M$ is a regular neighborhood for $X \cap \partial M$ in ∂M. For a proof of the following theorem see theorem 3.8 on page 88 in [12].

THEOREM 10.1. *For any compact P.L. subset $X \subset M$ there exists a regular neighborhood N for X in M such that N meets the boundary ∂M regularly. Moreover if N_1, N_2 are two regular neighborhoods for X in M which meet ∂M regularly then there is a P.L. homeomorphism $h : M \to M$ which maps N_1 onto N_2.*

REMARK 10.2 If in 10.1 we have that $N_1 \cap \partial M = N_2 \cap \partial M$, then we may assume that $h|\partial M = 1$ and that $h : M \to M$ is isotopic to the identity map mod $h|\partial M$.

There are the following useful facts which can be deduced from 10.1, 10.2, and the collaring theorem (see theorem 6.11 and corollary 6.12 in [12]).

FACT 10.3. Any regular neighborhood of any point in M is a P.L. cell of dimension equal m ($m = \dim(M)$).

FACT 10.4. Let e'_i, e_i be P.L. cells of dimension $n-1, n$ with $e'_i \subset \partial e_i$, for $i = 1, 2$. Any P.L. homeomorphism $h : e'_1 \to e'_2$ extends to a P.L. homeomorphism $h : e_1 \to e_2$. Moreover any P.L. homeomorphism $h : \partial e_1 \to \partial e_2$ extends to a P.L. homeomorphism $h : e_1 \to e_2$.

FACT 10.5. Let $A \subset M$ denote a codimension zero compact P.L. submanifold of M such that $A \cap \partial M$ is a compact codimension zero P.L. submanifold of ∂M. Let K denote a P.L. triangulation for M which also triangulates A, and let $K^{(2)}$ be a second derived subdivision of K. Set N equal the subcomplex of $K^{(2)}$ whose underlying set is the union of all closed simplices in $K^{(2)}$ which intersect B, where B is a given subcomplex of K lying in ∂M. (That is, $N = \text{star}(B; K^{(2)})$.) Then the following properties hold.

(a) There is a P.L. homeomorphism $h : (A, \partial_+ A) \to (\overline{A - N}, \overline{\partial_+ A - N})$, where $\partial_+ A = \overline{\partial A - A \cap \partial M}$.

(b) If A is a P.L. m-cell and $A \cap \partial M$ is a P.L. $(m-1)$-cell, then $\partial_+ A$ is a P.L. $(m-1)$-cell.

(c) If M, A are P.L. m-cells and $\partial_+ A$ is a P.L. $(m-1)$-cell, then $\overline{M - A}$ is a P.L. m-cell.

In the following lemma $(M, \partial M)$ will be a P.L. manifold pair and $\{(A_i, \partial A_i) : i \in I\}$ will denote a finite set of compact P.L. manifold pairs satisfying all the following properties.

10.6. (a) $\dim(A_i) = m$ (where $m = \dim(M)$), and A_i is a P.L. subset of $M - \partial M$ for all $i \in I$.

(b) If for some $I' \subset I$ we have $\cap_{i \in I'} \partial A_i \neq \emptyset$ then the cardinality of I' is less than $\dim(M) + 1$. Moreover there is an ordering $i_1, i_2, \ldots, i_k$ for the indices in I' and a P.L. embedding $g : (\cap_{i \in I'} \partial A_i) \times R^k \to M$ satisfying (c).

(c) For each subset $J \subset \{1, 2, \ldots, k\}$ set $R^k_J = \{(x_1, x_2, \ldots, x_k) \in R^k : x_i = 0 \text{ if } i \in J\}$. Then we must have that

$$g\left((\cap_{i \in I'} \partial A_i) \times R^k_J\right) = g\left((\cap_{i \in I'} \partial A_i) \times R^k\right) \cap \left(\cap_{j \in J} \partial A_j\right).$$

Recall (see page 50 of [12]) that for a simplicial complex K and a subcomplex $L \subset K$ the **derived neighborhood** for L in K—denoted by $N(L; K)$—is the union of all closed simplices in K which intersect L. An rth derived neighborhood for L in K—denoted by $N(L^{(r)}; K^{(r)})$—is the derived neighborhood for $L^{(r)}$ in $K^{(r)}$, where $K^{(r)}$ is an r-fold derived subdivision of K and $L^{(r)}$ is the subcomplex of $K^{(r)}$ satisfying $|L^{(r)}| = |L|$.

LEMMA 10.7. *Let K denote a P.L. triangulation for M which triangulates each A_i, $i \in I$. Let L be a subcomplex of K contained in ∂A_i for some $i \in I$. Let $N(L^{(2)}; K^{(2)})$ be a second derived neighborhood for L in K, and set $A'_i = A_i \cup N(L^{(2)}; K^{(2)})$.*

(a) There is a P.L. homeomorphism $h : M \to M$ which maps A_j onto itself for each $j \in I - i$ and which maps A_i onto A'_i.

(b) Given any neighborhood U for $N(L^{(2)}; K^{(2)})$ in M we may assume that $h|(M - U) = \text{identity}$.

REMARK 10.7. Lemma 10.7 is also true in the P.S. category. That is if "P.L." is replaced everywhere in lemma 10.7 by "P.S." then the resulting lemma is correct. This P.S. version is an immediate corollary of the P.L. version.

PROOF OF LEMMA 10.7. We divide the proof of this lemma into the following three steps.

STEP I. Let B denote the partition of M by maximal non-empty intersections of sets in the collection $\{\partial A_i, M - A_i, A_i - \partial A_i : i \in I\}$. Note it follows from 10.6 that B is a sort of cell structure for M in that the following properties hold.

10.8. For each $e \in B$ the closure $\overline{e}$ is a compact P.L. manifold with boundary $\partial\overline{e}$; for $e, e' \in B$ each of $\partial\overline{e}, \overline{e} \cap \overline{e'}$ is a union of a finite number of members of B; $e \cap e' = \emptyset$ if $e \neq e'$.

Note it follows from 10.6, and from corollary 6.12 in [12], that there is a collar $c : \partial A_i \times [0,1] \to \overline{M - A_i}$ for ∂A_i in $\overline{M - A_i}$ satisfying the following property.

10.9. (a) $c(a \times 0) = a$ for all $a \in \partial A_i$.

(b) For each $e \in B$ with $e \subset \partial A_i$ let $e(i)$ denote the unique member of B satisfying $\partial A_i \cap \overline{e(i)} = \overline{e}$, $e(i) \subset M - A_i$, and $\dim(e(i)) = \dim(e) + 1$. Then $c(e \times (0,1]) \subset e(i)$.

Let N' denote the image under $c : \partial A_i \times [0,1] \to \overline{M - A_i}$ of the set $(N(L^{(2)}; K^{(2)}) \cap \partial A_i) \times [0,1]$. Set $A_i'' = A_i \cup N'$. In this step we show that the following is true.

CLAIM 10.10. There is a P.L. homeomorphism $r : A_i' \to A_i''$ satisfying the following two properties.

(a) $r|A_i = 1$.

(b) For every $j \in I - i$ we have $r(A_i' \cap A_j) = A_i'' \cap A_j$ and $r(A_i' \cap \partial A_j) = A_i'' \cap \partial A_j$.

Towards verifying 10.10 we first note that for any $e, e(i)$ as in 10.9(b) both $N(L^{(2)}; K^{(2)}) \cap \overline{e(i)}$ and $N' \cap \overline{e(i)}$ are regular neighborhoods for $L \cap \overline{e(i)}$ in $\overline{e(i)}$ which meet $\partial\overline{e(i)}$ regularly. Because $L \cap \overline{e(i)} = L \cap \partial\overline{e(i)}$ we can get a third and fourth such regular neighborhoods X_e, Y_e for $L \cap \overline{e(i)}$ in $\overline{e(i)}$ by choosing a P.L. collaring $d : \partial\overline{e(i)} \times [0,1] \to \overline{e(i)}$ for $\partial\overline{e(i)}$ in $\overline{e(i)}$ and setting X_e (or Y_e) equal the image of $(N(L^{(2)}; K^{(2)})) \cap \partial\overline{e(i)} \times [0,1]$ (or of $(N' \cap \partial\overline{e(i)}) \times [0,1]$) under $d : \partial\overline{e(i)} \times [0,1] \to \overline{e(i)}$. Note that X_e and Y_e also meet the boundary $\partial\overline{e(i)}$ regularly. So we may apply theorem 10.1, 10.2 to get homeomorphisms $c_e^1 : \overline{e(i)} \to \overline{e(i)}$ and $c_e^2 : \overline{e(i)} \to \overline{e(i)}$ that satisfy the following properties.

10.11. (a) $c_e^1(X_e) = N(L^{(2)}; K^{(2)}) \cap \overline{e(i)}$ and $c_e^2(Y_e) = N' \cap \overline{e(i)}$.
(b) $c_e^1 | \partial \overline{e(i)} = id = c_e^2 | \partial \overline{e(i)}$.

It is left as an exercise to show that there exists a homeomorphism $r : A_i' \to A_i''$ uniquely determined by the following properties.

10.12 (a) $r | A_i = 1$.
(b) For any $e \in B$ with $e \subset \partial A_i$, and c_e^1, c_e^2 as in 10.11 we must have $c_e^2(r(x), t) = r \circ c_e^1(x, t)$ for all $t \in [0,1]$ and all $x \in N(L^{(2)}; K^{(2)}) \cap \partial \overline{e(i)}$
.

Now it follows from 10.11 that $r : A_i' \to A_i''$ defned by 10.12 satisfies claim 10.10.

This completes step I.

STEP II. In this step we construct a P.L. isotopy $\psi : A_i \times [0,1] \to M$ satisfying the following properties.

10.13. (a) $\psi(x, 0) = x$ for all $x \in A_i$.
(b) $\psi(A_i \times t) \subset A_i''$ for all $t \in [0,1]$.
(c) $\psi(A_i \times 1) = A_i''$.
(d) For any $j \in I - i$ and any $t \in [0,1]$ we must have that $\psi((A_j \cap A_i) \times t) \subset A_j$ and $\psi((\partial A_j \cap A_i) \times t) \subset \partial A_j$.

Towards verifying 10.13 note that there is a P.L. collaring $c_i : (\partial N(L^{(2)}; K^{(2)}) \cap \partial A_i) \times [0,1] \to N(L^{(2)}; K^{(2)}) \cap \partial A_i$ satisfying the following (see corollary 6.12 in [12], and 10.6 above).

10.14. $c_i((X \cap \partial A_i) \times [0,1]) \subset Y \cap \partial A_i$, where Y may be any of the sets $A_j \cap N(L^{(2)}; K^{(2)})$ or $\partial A_j \cap N(L^{(2)}; K^{(2)})$ with $j \in I - i$, and $X = Y \cap \partial N(L^{(2)}; K^{(2)})$.

Extend the collaring $c : A_i \times [0,1] \to \overline{M - A_i}$ of 10.9 to a bicollaring $C : \partial A_i \times [-1,1] \to M$ satisfying the following.

10.15. (a) $C | \partial A_i \times [0,1] = c$.
(b) $C((X \cap \partial A_i) \times [-1,1]) \subset X \cap A_i$, where X may be any of the sets $A_j, \partial A_j$ with $j \in I - i$.

Now set $S = [0,1] \times [-1,1]$ and $S^- = [0,1] \times [-1,0]$ and choose a P.L. isotopy $\phi : S^- \times [0,1] \to S$ satisfying the following.

10.16 (a) $\phi(\partial_1 S^- \times 1) = \partial_1 S$, where $\partial_1 S^- = [0,1] \times 0$ and $\partial_1 S = 0 \times [0,1] \cup [0,1] \times 1$.
(b) $\phi(x,t) = x$ for any $x \in \partial_2 S^-$ and any $t \in [0,1]$, where $\partial_2 S^- = 0 \times [-1,0] \cup [0,1] \times (-1)$.
(c) For any $(1,s) \in 1 \times [-1,0]$ and any $t \in [0,1]$ we have $\phi((1,s),t) = (1,(1+t)s+t)$.
(d) $\phi(x,0) = x$ for all $x \in S^-$.

Finally we can define $\psi : A_i \times [0,1] \to M$ as follows.

10.17. (a) For any $x \in A_i - C((N(L^{(2)}; K^{(2)}) \cap \partial A_i) \times [-1,0])$ and any $t \in [0,1]$ set $\psi(x,t) = x$.

(b) For any $x \in (N(L^{(2)}; K^{(2)}) \cap \partial A_i) - \text{image}(c_i)$ and any $s \in [-1,0]$ and $t \in [0,1]$ set $\psi(C(x,s),t) = C(x,(1+t)s+t)$.

(c) For any $y \in \partial(N(L^{(2)}; K^{(2)}) \cap \partial A_i)$ and any $s_1 \in [0,1]$, $s_2 \in [-1,0]$ and $t \in [0,1]$ set $\psi(C(c_i(y,s_1),s_2),t) = C(c_i(y,s_1'),s_2')$, where $(s_1',s_2') = \phi((s_1,s_2),t)$.

It is left as an exercise to deduce from 10.14-10.16 that $\psi : A_i \times [0,1] \to M$ defined by 10.17 is a well defined P.L. isotopy satisfying 10.13.

This completes step II.

STEP III. In this step we complete the proof of lemma 10.7. We denote by B_i the partitioning of M by all non-empty maximal intersections of sets in the collection $\{\partial A_j, A_j - \partial A_j, M - A_j : j \in I - i\}$. Set $\psi' : A_i \times [0,1] \to A_i'$ equal to the composite of maps

$$A_i \times [0,1] \xrightarrow{\psi} A_i'' \xrightarrow{r^{-1}} A_i'$$

where ψ comes from 10.13 and r comes from 10.10. Note it follows from 10.10, 10.13 that the following holds.

10.18. (a) $\psi'(x,0) = x$ for all $x \in A_i$.

(b) $\psi'(A_i \times 1) = A_i'$,

(c) For any $e \in B_i$ we must have $\psi'((e \cap A_i) \times t) \subset e \cap A_i'$ for all $t \in [0,1]$.

(d) $\psi' : A_i \times [0,1] \to M$ is a P.L. isotopy.

Now 10.18 makes possible the inductive application of the P.L. isotopy extension theorem (see theorem 6.12 in [12]) to each $e \in B_i$—the induction proceeds over dimension(e)—so as to extend the P.L. isotopy $\psi' : A_i \times [0,1] \to M$ of 10.18 to a P.L. isotopy $\overline{\psi} : M \times [0,1] \to M$ satisfying the following properties.

10.19. (a) $\overline{\psi}(x,0) = x$ for all $x \in M$.

(b) $\overline{\psi} = \psi'$ on $A_i \times [0,1]$.

(c) If $e \in B_i$ and $e \cap A_i = \emptyset$ then $\overline{\psi}(x,t) = x$ for all $x \in e$ and $t \in [0,1]$.

(d) $\overline{\psi}(e \times t) \subset e$ for all $e \in B_i$ and all $t \in [0,1]$.

Finally set $h : M \to M$ equal to the composite $M = M \times 1 \subset M \times [0,1] \xrightarrow{\overline{\psi}} M$. It follows from 10.18, 10.19 that $h : M \to M$ is a P.L. homeomorphism which satisfies the conclusions of lemma 10.7.

This completes the proof of lemma 10.7.

In the next lemma $(M, \partial M)$ denotes a compact P.L. manifold pair and $\{(A_i, \partial A_i) : i \in I\}$ denotes a finite set of compact P.L. manifold pairs which satisfy the following properties.

<u>10.20.</u> (a) Each A_i is a P.L. subset of M with $\dim(A_i) = \dim(M)$.

(b) For each $p \in M$ there is a minimal A_i in the collection $\{A_j : j \in I\}$ such that $p \in \mathrm{Int}(A_i)$. (We call A_i minimal if $A_j - A_i \neq \emptyset$ for all $j \in I - i$.)

(c) For each A_i set $\partial_1 A_i = A_i \cap \partial M$, $\partial_2 A_i = \overline{\partial A_i - \partial_1 A_i}$, $\Lambda A_i = \partial_1 A_i \cap \partial_2 A_i$. Then $\partial_1 A_i$, $\partial_2 A_i$ are compact codimension zero P.L. submanifolds of ∂A_i, and $\partial\partial_1 A_i = \partial\partial_2 A_i = \Lambda A_i$. For any $J \subset I$ and $J' \subset I - J$ set $Y(J, J') = (\bigcap_{j \in J} \partial_2 A_j) \cap (\bigcap_{j \in J'} A_j)$. Suppose that for every $i \in J$, $j \in J'$ we have that $A_j - A_i \neq \emptyset$ and $A_i - A_j \neq \emptyset$, and also suppose that $J' \neq \emptyset$. Then $Y(J, J')$ is either the empty set or is a P.L. $(m - \ell)$-cell, where $\ell = |J|$ and $m = \dim M$.

(d) All of the $\{(A_i, \partial_1 A_i) : i \in I\}$ are in P.L. transverse position to one another in $(M, \partial M)$.

Let B denote the partitioning of M by the non-empty maximal intersections of sets in the collection $\{A_i, \overline{M - A_i}, \partial_2 A_i : i \in I\}$.

LEMMA 10.21. *Each $e \in B$ is a P.L. cell.*

REMARK 10.21. Lemma 10.21 is also true in the P.S. category. That is, if "P.L." is replaced by "P.S." in 10.20 and lemma 10.21, then the resulting lemma is true. This P.S. version of lemma 10.21 is an immediate corollary of the P.L. version.

PROOF OF LEMMA 10.21. The proof is divided into the following two steps.

STEP I. Let J_1, J_2, J_3 denote subsets of the index set I which satisfy the following properties.

<u>10.22.</u> (a) J_1, J_2, J_3 are pairwise disjoint.

(b) For any $i \in J_1$, $j \in J_2$, and $k \in J_3$ we must have that all the following inequalities hold: $A_i - A_j \neq \emptyset$; $A_j - A_i \neq \emptyset$; $A_i - A_k \neq \emptyset$; $A_k - A_i \neq \emptyset$; $A_j - A_k \neq \emptyset$.

(c) There is $a \in J_2$ such that $A_k - A_a \neq \emptyset$ for all $k \in J_3$.

(d) For $i = 1, 2$, or 3 and any $x, y \in J_i$ we have $A_x - A_y \neq \emptyset$.

In this step we verify the following claim.

CLAIM 10.23. For J_1, J_2, J_3 as in 10.22 the following must hold.

(a) The set $Y(J_1, J_2, J_3)$, which is defined to be the intersection $(\bigcap_{j \in J_1} \partial_2 A_j) \cap (\bigcap_{j \in J_2} A_j) \cap (\bigcap_{j \in J_3} (\overline{M - A_j}))$, must be either the empty set or P.L. $(m - \ell)$-cell, where $\ell = |J_1|$.

To verify 10.23(a) we proceed by induction over the cardinality of J_3 (denoted by $|J_3|$). First note that if $|J_3| = 0$ then 10.23(a) follows from 10.20(c). We now assume that 10.23(a) is true for all J_3 with $|J_3| \leq \ell$, and show that 10.23(a) is also true for J_3 with $|J_3| = \ell + 1$.

To carry out this induction step we consider the following subsets of I. Fix an element $k \in J_3$ and set

10.24.
$$\begin{aligned} J_1' &= J_1 \cup \{k\}, \\ J_2' &= \{j : j \in J_2,\ A_k - A_j \neq \emptyset\}, \\ J_2'' &= J_2' \cup \{k\}, \\ J_3' &= J_3 - \{k\}. \end{aligned}$$

Note that all of the triples (J_1', J_2', J_3'), (J_1, J_2'', J_3') and (J_1, J_2, J_3') all satisfy 10.22, and $|J_3'| = \ell$. So by induction the sets $Y(J_1', J_2', J_3')$, $Y(J_1, J_2'', J_3')$ and $Y(J_1, J_2, J_3')$ are either empty sets of P.L. cells of dimensions $m - |J_1| - 1$, $m - |J_1|$, and $m - |J_1|$ respectively. If $Y(J_1', J_2', J_3') = \emptyset$ then either $Y(J_1, J_2, J_3)$ is equal to the empty set or to $Y(J_1, J_2, J_3')$, which verifies the induction step. On the other hand if $Y(J_1', J_2', J_3') \neq \emptyset$ then to verify the induction step we need to consider the following sets:

10.25. $Y = [\partial Y(J_1, J_2, J_3') \times [0,1]] \cup Y(J_1, J_2, J_3')$, where the union is taken along the identification $\partial Y(J_1, J_2, J_3') \times 0 = \partial Y(J_1, J_2, J_3')$; $Y_1 = \partial Y(J_1, J_2, J_3') \times [0,1]$; $Y_2 = Y_1 \cup Y(J_1, J_2'', J_3')$, where the union is taken along the identification $Y(J_1, J_2'', J_3') \cap \partial Y(J_1, J_2, J_3') = Y(J_1, J_2'', J_3') \cap Y_1$. (Note $Y(J_1, J_2'', J_3') \subset Y(J_1, J_2, J_3')$.)

Note that Y_1 is a regular neighborhood of ∂Y in Y. Note also that there is an elementary collapse $Y_2 \searrow Y_1$ (collapse $Y(J_1, J_2'', J_3')$ across the ball $Y(J_1', J_2', J_3')$), and that Y_2 is a P.L. manifold neighborhood for ∂Y in Y. It follows that Y_2 is also a regular neighborhood for ∂Y in Y. So by theorem 10.1 there is a P.L. homeomorphism $h : Y \to Y$ such that h maps $\overline{Y - Y_1}$ onto $\overline{Y - Y_2}$. But $\overline{Y - Y_1} = Y(J_1, J_2, J_3')$, which is a P.L. cell of dimension $m - |J_1|$ (by induction), and $\overline{Y - Y_2} = Y(J_1, J_2, J_3)$. So $Y(J_1, J_2, J_3)$ is homeomorphic to the P.L. $(m - |J_1|)-$ cell $Y(J_1, J_2, J_3')$ as required.

This completes the verification of 10.23(a).

STEP II. In this step we complete the proof of lemma 10.21.

Note for any $e \in B$ there are subsets J_1, J_2, J_3 of I satisfying the following properties (see 10.20(b)).

10.26. (a) J_1, J_2, J_3 satisfy 10.22(a)(b)(c)(d).
(b) $e = Y(J_1, J_2, J_3)$.

Now the conclusion of lemma 10.21 follows from 10.26 and claim 10.23.
This completes the proof of lemma 10.21.

§11. PROOF OF THE THICKENING THEOREM

In this section we prove theorem 9.6.

There will be no loss of generality in assuming that the following properties are satisfied by the $C, \{K_{u,e} : e \in C\}$, etc. These properties will be assumed to hold through section 13, and will be cited in sections 12, 13.

11.0. (a) Let q, U be as in 6.1. Then we have the following inclusions:

$$F^q(F^q(U) \cap (U - |C|)) \subset M - U;$$
$$F^{-q}(F^{-q}(U) \cap (U - |C|)) \subset M - U;$$
$$\bigcup_{e \in C} g_e(X_{u,e} \times X_{s,e}) \subset U.$$

(b) Set $X = \cup_{e \in C}\, g_e({}_0N_{u,e} \times {}_0N_{s,e})$, $X_1 = \cup_{e \in C}\, g_e({}_1N_{u,e} \times {}_1N_{s,e})$, where ${}_0N_{u,e}$, ${}_0N_{s,e}$, ${}_1N_{u,e}$, ${}_1N_{s,e}$ come from 8.1, 8.2, 8.4, 8.5, 9.0. Let $\partial|C|$, ∂X, $\partial X_1, \partial\overline{U}$ denote the topological boundaries of $|C|$, X, X_1, $\overline{U}$ in M, and let $\delta > 0$ denote the minimum distance between any two of the $\partial|C|$, ∂X, $\partial X_1, \partial\overline{U}$. We require that

$$a\lambda^q\eta^2\mu^{-2}\varepsilon << \delta, \quad a\lambda^{-q} << \mu^2\eta^{-2}$$

where a, λ^q come from 5.1, η comes from 6.1(e), μ comes from 7.2(b), and ε is an upper bound for the diameter of each $\{K_{u,e} : e \in C\}$ and each $\{B_{s,k} : k \in I_e,\ e \in C\}$.

(c) For any $e \in C,\ k \in I_e,\ f \in K_{u,e},\ i \in \{1, 2, \ldots, y\}$ set ${}_kX_i(f) = F^q \circ g_e({}_kY_i(f) \times {}_2B_{s,k})$. Then if ${}_kX_i(f) \subset X$ there is $e' \in C$ such that ${}_kX_i(f) \subset g_{e'}({}_0N_{u,e'} \times {}_0N_{s,e'})$. If ${}_kX_i(f) \subset g_{e'}({}_0N_{u,e'} \times {}_0N_{s,e'})$ then there is $k' \in I_{e'}$ such that ${}_kX_i(f) \subset g_{e'}({}_0N_{u,e'} \times B_{s,k'})$.

(d) For any $e \in C,\ k \in I_e$ let $\hat{B}_{s,k}$ denote the collection of all subsets $p \subset R^m$ gotten as follows. For each $p \in \hat{B}_{s,k}$ there is $e' \in C,\ k' \in I_{e'}$ such that $g_{e'}(X_{u,e'} \times {}_2B_{s,k'}) \cap g_e(X_{u,e} \times {}_2B_{s,k}) \neq \emptyset$ and p equals the projection of $g_{e'}(X_{u,e'} \times B_{s,k'}) \cap g_e(R^n \times R^m)$ to R^m. Set $\check{B}_{s,k} = \cup_{p \in \hat{B}_{s,k}}\, p$. Suppose that for $\omega \in \overline{X}_{u,e}$ and an integer $j \geq 0$ we have that $F^{qi} o g_e(\omega \times \check{B}_{s,k}) \subset U$ holds for all $i \in \{0, 1, 2, \ldots, j\}$. Then if for some $e_1, e_2 \in C$ we have that

$$F^{qj}(\omega \times \check{B}_{s,k}) \cap g_{e_\ell}(\overline{X_{u,e_\ell} \times X_{s,e_\ell}}) \neq \emptyset,$$

for $\ell = 1, 2$, then we must have that either $e_1 \subset e_2$ or $e_2 \subset e_1$.

REMARK. That 11.0(a) can be satisfied follows from 6.1(c), 6.2(b). That 11.0(b)(c) can be satisfied follows by choosing q large enough, and choosing the diameter of each $K_{u,e}$ and of each $B_{s,k}$ small enough, and then using a Lebesque covering argument (compare with 6.8(e), 8.1(a),(d), and 8.8(b)). That 11.0(d) can be satisfied follows from 5.1(b), 6.7(b) by choosing the diameter of each $B_{s,k}$ sufficiently small.

In the rest of this section e will denote a fixed member of C, k a fixed member of I_e, and j a positive integer. We define a set of smooth triangulations $\{K_{i,j} : 0 \leq i \leq n+m\}$ in $X_{u,e}$ (for each positive integer j) which depend only on the given e, k, and j.

First define a subset ${}_jX_{u,e} \subset X_{u,e}$ as follows.

11.1. (a) $y \in {}_jX_{u,e}$ if and only if

$$F^{qs} \circ g_e((y) \times \check{B}_{s,k}) \subset U \quad \text{holds for all } s \in \{0,1,2,\ldots,j\}.$$

Now for each $i \in \{0,1,2,\ldots,n+m\}$ define a smooth triangulation $L_{i,j}$ in ${}_jX_{u,e}$ as follows.

11.1. (b) A smooth simplex $\Delta \subset {}_jX_{u,e}$ is in $L_{i,j}$ if and only if there is $e' \in C$, $\Delta' \in \tilde{K}_{u,e'}$, and $k' \in \tilde{I}_{e'}$, so that $\dim(e') = i$, and

$$F^{qj} \circ g_e(\Delta \times p) = g_{e'}(\Delta' \times {}_{3/2}B_{s,k'}) \cap F^{qj} \circ g_e(Y \times p),$$

for some $p \in \hat{B}_{s,k}$ and some neighborhood Y of Δ. (Here $\ell \in \tilde{I}_{e'}$ if and only if the distance from $B_{s,\ell}$ to ${}_0N_{s,e'}$ is less than $\delta\eta^{-2}/2$; and $\tilde{K}_{u,e'}$ is the subcomplex of all simplices in $K_{u,e'}$ a distance less than $\delta\eta^{-2}/2$ from ${}_0N_{u,e'}$ (see 11.0(b) for δ, η).)

Let L be a finite smooth triangulation in R^n, and denote by $\partial|L|$ the topological boundary of $|L|$ in R^n. We let ${}_3L$ denote the maximal subcomplex of L such that for any three simplices $\Delta_1, \Delta_2, \Delta_3 \in L$ if $\Delta_1 \cap {}_3L \neq \emptyset$, $\Delta_1 \cap \Delta_2 \neq \emptyset$, $\Delta_2 \cap \Delta_3 \neq \emptyset$, then $\Delta_3 \cap \partial L = \emptyset$.

We can now define the $K_{i,j}$ by induction as follows.

11.1. (c) $K_{0,j} = {}_3L_{0,j}$.

(d) $K_{i,j}$ is the maximal subcomplex of ${}_3L_{i,j}$ satisfying: for any $i' < i$, and any $\Delta \in K_{i',j}$, if $(\Delta - \partial\Delta) \cap |K_{i,j}| \neq \emptyset$ then $|\Delta|$ is the underlying set of a subcomplex of $K_{i,j}$.

Note it follows from 6.8(d) and 11.0 that the $\{K_{i,j}\}$ are well defined by 11.1(c)(d). Note also that it follows from 11.0, 11.1, 6.8(d) that the $\{K_{i,j}\}$ satisfy the following properties.

11.2. (a) For any $i, i' \in \{0, 1, \ldots, n+m\}$ we have that $|K_{i,j}| \cap |K_{i',j}|$ is the underlying set of subcomplexes of both $K_{i,j}$ and $K_{i',j}$.

(b) If $i < i'$ then for any $\Delta \in K_{i,j}$ with $(\Delta - \partial\Delta) \cap |K_{i',j}| \neq \emptyset$ we have that Δ is the underlying set of a subcomplex of $K_{i',j}$.

(c) The $\{K_{i,j} : 0 \leq i \leq m+n\}$ depend on e, k, j.

For each $i \in \{0, 1, \ldots, n+m\}$ and each $\Delta \in K_{i,j}$ we define a set of P.S. balls $\{Y_\ell(\Delta) : 1 \leq \ell \leq \phi(\Delta)\}$ as follows: for each $Y_\ell(\Delta)$ there must be $e' \in C$, $k' \in \tilde{I}_{e'}$, $\Delta' \in \tilde{K}_{u,e'}$ as in 11.1(b), and a ball ${}_{k'}Y_{\ell'}(\Delta')$—as in 8.2—so that the following hold.

11.3. $F^{qj}(g_e(\Delta \times p)) = g_{e'}(\Delta' \times {}_{3/2}B_{s,k'}) \cap F^{qj}(g_e(Y \times p))$,
and

$$F^{qj}(g_e(Y_\ell(\Delta) \times p)) = g_{e'}({}_{k'}Y_{\ell'}(\Delta') \times {}_{3/2}B_{s,k'}) \cap F^{qj}(g_e(Y \times p)),$$

for some $p \in \hat{B}_{s,k}$ and some neighborhood Y for $\Delta \cup Y_\ell(\Delta)$.

Note it follows from 7.2, 7.1, 11.0, 11.1, that the $\{Y_\ell(\Delta) : \Delta \in K_{i,j},\ 1 \leq \ell \leq \phi(\Delta)\}$ satisfy the following properties.

11.4. (a) $Y_{\ell+1}(\Delta) \subset \mathrm{Int}(Y_\ell(\Delta))$ for any $\Delta \in K_{i,j}$, any $i \in \{0, 1, \ldots, n+m\}$, and any $\ell \in \{1, 2, \ldots, \phi(\Delta) - 1\}$. (This will be true after reordering the $\{Y_\ell(\Delta) : 1 \leq \ell \leq \phi(\Delta)\}$, if necessary.)

(b) Let J, J' be subsets of $\bigcup_{x=0}^{n+m} K_{x,j}$ such that $J \cap J' = \emptyset$, and for any $\Delta_1, \Delta_2 \in J \cup J'$ if $\Delta_1 \subset \Delta_2$ then either $\Delta_1 \subset \partial\Delta_2$ or $\Delta_1 = \Delta_2$. For each $\Delta' \in J \cup J'$ choose one of the $\{Y_\ell(\Delta') : 1 \leq \ell \leq \phi(\Delta')\}$ and denote it by $Y(\Delta')$. For any given $\Delta \in \bigcup_{x=0}^{n+m} K_{x,j}$ we have that

$$\Delta \bigcap \left(\bigcap_{\Delta' \in J} Y(\Delta') \right) \bigcap \left(\bigcap_{\Delta' \in J'} \partial Y(\Delta') \right) \neq \emptyset$$

if and only if there is an ordering—$\Delta_1, \Delta_2, \ldots, \Delta_p$—of the simplices in $J \cup J'$ such that the following hold.

(i) For each $1 \leq i < j \leq p$ we have $(\Delta_i - \partial\Delta_i) \cap \Delta_j \neq \emptyset$ and $\Delta_i \cap \Delta_j \subset \partial\Delta_j$. Also $e_{i'} \subset e_{j'}$, where $\Delta_i \in K_{u,e,e_{i'}}$, $\Delta_j \in K_{u,e,e_{j'}}$.

(ii) $(\Delta_i - \partial\Delta_i) \cap \Delta \neq \emptyset$ for all $1 \leq i \leq p$.

(iii) $\dim(\Delta) - |J'| \geq 0$, where $|J'|$ is the cardinality of J'.

(c) For $\Delta_1, \Delta_2 \in \cup_{x=0}^{n+m} K_{x,j}$ suppose that $\Delta_1 \subset \Delta_2$, $\Delta_1 \not\subset \partial\Delta_2$, $\Delta_1 \neq \Delta_2$. If $\Delta_1 \subset \Delta_2 - \partial\Delta_2$, we require that $Y_\ell(\Delta_1) \subset \mathrm{Int}(Y_k(\Delta_2))$ for all $1 \leq \ell \leq \phi(\Delta_1)$ and all $1 \leq k \leq \phi(\Delta_2)$. Even if $\Delta_1 \not\subset \Delta_2 - \partial\Delta_2$, we require that $\cup_{\Delta' \in \Delta_1} Y(\Delta') \subset \mathrm{Int}(\cup_{\Delta'' \in \Delta_2} Y(\Delta''))$, where $Y(\Delta')$, $Y(\Delta'')$ can be any of $\{Y_\ell(\Delta') : 1 \leq \ell \leq \phi(\Delta')\}$, $\{Y_k(\Delta'') : 1 \leq k \leq \phi(\Delta'')\}$.

(d) Suppose that J, J', Δ are as in (b) and if $J = \emptyset$ then $\Delta \notin J'$. Then any non-empty intersection $\Delta \cap (\cap_{\Delta' \in J} Y(\Delta')) \cap (\cap_{\Delta' \in J'} \partial Y(\Delta'))$ of part (b) is a P.S. cell of dimension equal $n' - |J'|$, where $n' = \dim(\Delta)$. Moreover, if $J \neq \emptyset$, then any non-empty intersection $\partial\Delta \cap (\cap_{\Delta' \in J} Y(\Delta')) \cap (\cap_{\Delta' \in J'} \partial Y(\Delta'))$ is a P.S. cell of dimension equal $n' - |J'| - 1$. Finally if $J \neq \emptyset$ then $(\cap_{\Delta' \in J} Y(\Delta')) \cap (\cap_{\Delta' \in J'} \partial Y(\Delta'))$ is a P.S. cell of dimension equal $n - |J'|$ if it is not empty.

(e) For any $p \in \cup_{i=0}^{n+m} |K_{i,j}|$ there is a minimal $Y_\ell(\Delta)$ in the collection $\{Y_{\ell'}(\Delta') : \Delta' \in K_{i,j},\ 1 \leq \ell' \leq \phi(\Delta'),\ 0 \leq i \leq n+m\}$ such that $p \in \mathrm{Int}(Y_\ell(\Delta))$. (Recall $Y_\ell(\Delta)$ is minimal if for every other $Y_{\ell'}(\Delta')$ we have $Y_{\ell'}(\Delta') - Y_\ell(\Delta) \neq \emptyset$.)

We also need to define for each $i \in \{0, 1, \ldots, n+m\}$ and each $\Delta \in K_{i,j}$ P. S. balls $\{Y'_\ell(\Delta),\ Y_\ell(\Delta; 1) : 1 \leq \ell \leq \phi(\Delta)\}$ in R^n as follows: for each $Y'_\ell(\Delta)$ and $Y_\ell(\Delta; 1)$ there must be $e' \in C$, $k' \in \tilde{I}_{e'}$, and $\Delta' \in \tilde{K}_{u,e'}$, and ${}_{k'}Y_{\ell'}(\Delta')$ as in 11.3 such that

$$F^{qj}(g_e(Y'_\ell(\Delta) \times p)) = g_{e'}({}_{k'}Y'_{\ell'}(\Delta') \times {}_{3/2}B_{s,k'}) \cap F^{qj}(g_e(Y \times p)),$$

and

$$F^{qj}(g_e(Y_\ell(\Delta; 1) \times p)) = g_{e'}({}_{k'}Y(\Delta'; 1) \times {}_{3/2}B_{s,k'}) \cap F^{qj}(g_e(Y \times p)),$$

for some $p \in \hat{B}_{s,k}$, and for some neighborhood Y for $Y_\ell(\Delta; 1)$.

Here ${}_{k'}Y_{\ell'}(\Delta')$ and ${}_{k'}Y_{\ell'}(\Delta'; 1)$ are as in 8.5 and 9.5. Note that the following properties hold for the $Y'_\ell(\Delta)$ and $Y_\ell(\Delta; 1)$ (see 7.2, 8.2, 8.5, 9.2, 9.5, 11.0, 11.1).

<u>11.5</u> (a) Let $\{A_i : i \in I\}$ of 10.6 be defined to be the $\{Y'_\ell(\Delta) : \Delta \in \cup_{x=0}^{n+m} K_{x,j-1},\ 1 \leq \ell \leq \phi(\Delta)\}$, and let M of 10.6 be defined to be R^n. Then these $\{A_i : i \in I\}$ and M satisfy the P.S. versions of 10.6(a)(b)(c).

(b) For each A_i and each $K_{y,j}$, we have that $A_i \cap |K_{y,j}|$ is the underlying set of a subcomplex of $K_{y,j}$. (Compare with theorem 8.5(c).)

(c) For each A_i either $Y'_\ell(\Delta) = Y_\ell(\Delta; 1)$—where $A_i = Y'_\ell(\Delta)$—or there is a smooth triangulation $K_{y,j}$ with $A_i \subset \mathrm{Int}(|K_{y,j}|)$, and for each $\Delta \in K_{y,j}$ with $\Delta \in \partial A_i$ there is a ball in $\{Y'_\ell(\Delta) : 1 \leq \ell \leq \phi(\Delta)\}$—denoted by $Y(A_i, \Delta)$. For all other $\Delta' \in \cup_{x=0}^{n+m} K_{x,j}$ we set $Y(A_i, \Delta') = \emptyset$. The collection of all these $\{Y(A_i, \Delta) : i \in I,\ \Delta \in \cup_{x=0}^{n+m} K_{x,j}\}$ satisfy (d)(e).

(d) If $Y'_\ell(\Delta) \neq Y_\ell(\Delta; 1)$ we have that $Y_\ell(\Delta; 1) = A_i \cup (\cup Y(A_i, \Delta'))$, where the union $\cup Y(A_i, \Delta')$ runs over all $\Delta' \in \cup_{x=0}^{n+m} K_{x,j}$ and $Y'_\ell(\Delta) = A_i$.

(e) If $\Delta \subset \partial A_i \cap \partial A_{i'}$ for some $\Delta \in \cup_{x=0}^{n+m} K_{x,j}$, and $i \neq i'$, then $Y(A_i, \Delta) \neq Y(A_{i'}, \Delta)$ if one of the sets $Y(A_i, \Delta)$, $Y(A_{i'}, \Delta)$ is not empty.

(f) Suppose $A_i = Y'_\ell(\Delta)$ and $Y_\ell(\Delta; 1) = Y'_\ell(\Delta)$. Then we must have that $A_i \subset R^n - {}_qX_{u,e}$ holds for all $q > j$ (see 11.1(a) for ${}_qX_{u,e}$).

(g) Suppose that for $A_i = Y'_\ell(\Delta)$ we have that $Y_\ell(\Delta; 1) = Y'_\ell(\Delta)$. Let $A_{i_1}, A_{i_2}, A_{i_3}$ be such that $A_i \cap A_{i_1} \neq \emptyset$, $A_{i_1} \cap A_{i_2} \neq \emptyset$, $A_{i_2} \cap A_{i_3} \neq \emptyset$; and let $\Delta_1, \Delta_2, \Delta_3 \in \cup_{x=0}^{n+m} K_{x,j}$, $1 \leq \ell_i \leq \phi(\Delta_i)$ for $i = 1, 2, 3$, be such that $Y'_{\ell_1}(\Delta_1) \cap A_{i_3} \neq \emptyset$, $Y'_{\ell_1}(\Delta_1) \cap Y'_{\ell_2}(\Delta_2) \neq \emptyset$, $Y'_{\ell_2}(\Delta_2) \cap Y'_{\ell_3}(\Delta_3) \neq \emptyset$. Then we must have $Y_{\ell_3}(\Delta_3) = Y'_{\ell_3}(\Delta_3) = Y_{\ell_3}(\Delta_3; 1)$.

Now the remainder of this proof of theorem 9.6 is carried out in the following three steps.

STEP I. In this step we construct for each $x \in \{0, 1, \ldots, n+m\}$, each $\Delta \in K_{x,j}$, and each $\ell \in \{1, 2, \ldots, \phi(\Delta)\}$ a P. S. ball $\hat{Y}_\ell(\Delta)$ such that the following properties are satisfied.

<u>11.6.</u> (a) All the $\{\hat{Y}_\ell(\Delta) : 0 \leq x \leq n+m,\ \Delta \in K_{x,j},\ 1 \leq \ell \leq \phi(\Delta)\}$ satisfy 11.4(a)-(e) (when the $Y_\ell(\Delta)$ of 11.4 are replaced by the $\hat{Y}_\ell(\Delta)$).

(b) There is an ordering—denoted by $X_1, X_2, X_3, \ldots, X_t$—of all the P.S. balls $\{\hat{Y}_\ell(\Delta) : 0 \leq x \leq n+m,\ \Delta \in K_{x,j},\ 1 \leq \ell \leq \phi(\Delta)\}$. For each $a \in \{1, 2, \ldots, t\}$ there is a smooth triangulation S_a for R^n, a subcomplex S'_a of S_a, and a second derived subdivision $S_a^{(2)}$ of S_a which satisfy (c)(d).

(c) Suppose $X_a = \hat{Y}_\ell(\Delta)$ for $a \in \{1, 2, \ldots, t\}$, $\Delta \in \cup_{x=0}^{n+m} K_{x,j}$, and $\ell \in \{1, 2, \ldots, \phi(\Delta)\}$. Then we have $|S'_a| \subset \Delta - \partial\Delta$ and $\mathrm{Star}(|S'_a|, S_a^{(2)}) = \hat{Y}_\ell(\Delta)$, where $\mathrm{Star}(|S'_a|, S_a^{(2)})$ denotes the collection of all closed simplices $\Delta' \in S_a^{(2)}$ such that $\Delta' \cap |S'_a| \neq \emptyset$.

(d) S_a triangulates each of the following subsets of R^n : each A_i of 11.5(a); each simplex in $\cup_{x=0}^{n+m} K_{x,j}$; each of the $X_1, X_2, \ldots, X_{a-1}$.

Towards verifying 11.6 we begin by noting that the partition of R^n generated by all the smooth polyhedra $\{K_{i,j} : 0 \leq i \leq n+m\}$, $\{Y_\ell(\Delta) : 1 \leq \ell \leq \phi(\Delta),\ \Delta \in K_{i,j-1},\ 0 \leq i \leq n+m\}$ is a smooth polyhedron in R^n—denoted by W (see 7.2(a), 11.5(b)). Choose a smooth triangulation T of R^n which triangulates each member of W. We choose a sequence $T^{(1)}, T^{(2)}, \ldots, T^{(y)}, T^{(y+1)}, \ldots$ of derived subdivisions of T such that for all $y > 0$ $T^{(y+1)}$ is a first derived subdivision of $T^{(y)}$.

Now for each $\Delta \in \cup_{x=0}^{n+m} K_{x,j}$ and each $\ell \in \{1, 2, \ldots, \phi(\Delta)\}$ choose two positive integers—denoted $\psi_1(\Delta, \ell)$ and $\psi_2(\Delta, \ell)$—such that the following properties hold.

11.7. (a) If for $\Delta, \Delta' \in \cup_{x=0}^{n+m} K_{x,j}$ we have that $(\Delta - \partial\Delta) \cap \Delta' \neq \emptyset$ and $\Delta' \not\subset \Delta$, and either Δ is a face of $\partial\Delta'$ or $\Delta \not\subset \Delta'$, then $\psi_2(\Delta, \ell) < \psi_1(\Delta', \ell')$ holds for any ℓ, ℓ'.

(b) If $\Delta \subset \Delta'$ but $\Delta \neq \Delta'$ and Δ is not a face of $\partial\Delta'$, then $\psi_2(\Delta', \ell') < \psi_1(\Delta, \ell)$ holds for all ℓ, ℓ'.

(c) $10 + \psi_1(\Delta, \ell) < \psi_2(\Delta, \ell)$ for all Δ, ℓ. Moreover if $(\Delta, \ell) \neq (\Delta', \ell')$ then $|\psi_i(\Delta, \ell) - \psi_j(\Delta', \ell')| \geq 10$ holds for $i, j \in \{1, 2\}$.

(d) $\psi_1(\Delta, \ell + 1) < \psi_1(\Delta, \ell)$ and $\psi_2(\Delta, \ell) < \psi_2(\Delta, \ell + 1)$ hold for all Δ, ℓ.

Note by 11.7(c) there is an ordering of all the pairs $\{(\Delta, \ell)) : \Delta \in \cup_{x=0}^{n+m} K_{x,j},\ 1 \leq \ell \leq \phi(\Delta)\}$—denoted by $P_1, P_2, \ldots, P_t$—defined as follows: $\psi_2(P_i) < \psi_2(P_{i'})$ if and only if $i < i'$. We can now define the $\hat{Y}_\ell(\Delta)$, X_i, S_i, S_i' and $S_i^{(2)}$ of 11.6.

11.8. (a) Suppose $P_i = (\Delta, \ell)$. Set $S_i = T^{(x)}, S_i^{(2)} = T^{(x+2)}$, where $x = \psi_2(\Delta, \ell)$.

(b) Set $N = \overline{\Delta - |\mathrm{Star}(\partial\Delta, T^{(y+2)})|}$, where $y = \psi_1(\Delta, \ell)$. Now let S_i' denote the subcomplex of S_i such that $|S_i'| = N$.

(c) $X_i = \hat{Y}_\ell(\Delta)$, where $\hat{Y}_\ell(\Delta)$ is determined from S_i' and $S_i^{(2)}$ as in 11.6(c).

It is left as an exercise to deduce from 11.7, 11.8 that 11.6(b)(c)(d) hold, and that the $\{\hat{Y}_\ell(\Delta)\}$ satisfy 11.4(a)(b)(c)(e). To complete the verification of 11.6(a) it remains to show that the $\{\hat{Y}_\ell(\Delta)\}$ also satisfy 11.4(d). We verify 11.4(d) by induction over the order of the set $J \cup J'$ of 11.4(d). Suppose 11.4(d) is satisfied whenever the order of $J \cup J'$ is less than or equal to $p - 1$. Now let $\Delta_1, \Delta_2, \ldots, \Delta_p$, be as in 11.4(b). There are the following two cases to consider.

CASE 1. Suppose that $\Delta_p \subset J$. Set $\Delta_{p+1} = \Delta_p \cap \Delta$, $\Delta_{p+2} = \partial\Delta_p \cap \Delta$. By induction we have that

$$\Delta_{p+1} \bigcap \left(\bigcap_{\Delta' \in J - \Delta_p} \hat{Y}(\Delta') \right) \bigcap \left(\bigcap_{\Delta' \in J'} \partial\hat{Y}(\Delta') \right) = Z_1$$

is a P.S. cell of dimension equal $\dim(\Delta_{p+1}) - |J'|$.

Note it also follows by induction that

$$\Delta \bigcap \left(\bigcap_{\Delta' \in J - \Delta_p} \hat{Y}(\Delta') \right) \bigcap \left(\bigcap_{\Delta' \in J'} \partial\hat{Y}(\Delta') \right) = Z_2$$

is a P.S. cell of dimension equal $\dim(\Delta) - |J'|$. Note also that it follows from 11.7, 11.8(a)(b)(c) that $T^{(x)}$ triangulates $Z_1, Z_2, \Delta_{p+1}, \Delta_{p+2}$ —where $x = \psi_1(\Delta_p, \ell)$ and $\hat{Y}_\ell(\Delta_p) = \hat{Y}(\Delta_p)$—and that the set

$$\Delta \bigcap \left(\bigcap_{\Delta' \in J} \hat{Y}(\Delta') \right) \bigcap \left(\bigcap_{\Delta' \in J'} \partial \hat{Y}(\Delta') \right) = Z_3$$

can be gotten as follows. Set $N = \overline{Z_1 - |\mathrm{star}(\Delta_{p+2}, T^{(x+2)})|}$. [Then by 10.5—as applied to the subset $Z_1 \subset \Delta_{p+1}$—we conclude that N is also a P.S. cell.] Set $y = \psi_2(\Delta_p, \ell)$, let H denote the subcomplex of $T^{(y)}$ having underlying set equal Z_2, and let $H^{(2)}$ denote the subcomplex of $T^{(y+2)}$ having underlying set equal to Z_2. Then we have

$$Z_3 = |\mathrm{Star}(N, H^{(2)})|.$$

Since $\mathrm{star}(N, H^{(2)})$ is a regular neighborhood for N in Z_2, (see [10]) and both N, Z_2 are P.S. cells, it follows that Z_3 is a P.S. smooth cell of the same dimension as Z_2 (see 10.3).

A similar argument shows that $\left(\cap_{\Delta' \in J} \hat{Y}(\Delta')\right) \cap \left(\cap_{\Delta' \in J'} \partial \hat{Y}(\Delta')\right)$ is a P.S. cell of dimension equal $n - |J'|$.

Now the collapsing arguments used in Cases 1,3 on pages 44–46 and pages 47–48 can be used to show that

$$Z_4 = \partial \Delta \bigcap \left(\bigcap_{\Delta' \in J} Y(\Delta') \right) \bigcap \left(\bigcap_{\Delta' \in J'} \partial Y(\Delta') \right)$$

is also a P.S. cell of dimension equal to $\dim(\Delta) - |J'| - 1$. The details are left to the reader.

This completes Case 1.

CASE 2. Suppose that $\Delta_p \in J'$. It will be most convenient to return to the situation of Case 1 above (i.e., $\Delta_p \in J$) and show that

$$Z_5 = \Delta \bigcap \left(\bigcap_{\Delta' \in J - \Delta_p} Y(\Delta') \right) \bigcap \left(\bigcap_{\Delta' \in J' \cup \{\Delta_p\}} \partial Y(\Delta') \right)$$

and

$$Z_6 = \partial \Delta \bigcap \left(\bigcap_{\Delta' \in J - \Delta_p} Y(\Delta') \right) \bigcap \left(\bigcap_{\Delta' \in J' \cup \{\Delta_p\}} \partial Y(\Delta') \right)$$

are P.S. cells of dimensions equal $\dim(\Delta)-|J'|-1$ and $\dim(\Delta)-|J'|-2$.

We first consider Z_5. If $\Delta \subset \Delta_p$ note that Z_2 and $\Delta_{p+2} \cap Z_2$ are both P.S. cells (by Case 1). Note also that $N_1 = |\text{star}(\Delta_{p+2} \cap Z_2, T^{(x+2)})| \cap Z_2$ and $N_2 = N_1 \cap \partial Z_2$ are regular neighborhoods for $\Delta_{p+2} \cap Z_2$ in Z_2 and in ∂Z_2, and are therefore P.S. cells of dimension equal $\dim(Z_2)$ and $\dim(Z_2)-1$. (Here $x = \psi_1(\Delta_p, \ell)$ and $\hat{Y}_\ell(\Delta_p) = \hat{Y}(\Delta_p)$.) It follows from 10.5 that $N_3 = \overline{\partial N_1 - N_2}$ is a P.S. cell of dimension equal $\dim(Z_2) - 1$ and with $N_3 \cap \partial Z_2 = \partial N_3$. Now set $N_4 = |\text{star}(N_3, T^{(y+2)})| \cap N_1$ and set $N_5 = N_4 \cap N_2$. (Here $y = \psi_2(\Delta_p, \ell)$). Note that (N_4, N_5) is a regular neighborhood for $(N_3, \partial N_3)$ in (N_1, N_2). It follows that $N_6 = \overline{\partial N_4 - N_5 - N_3}$ is P.S. homeomorphic to N_3. Thus N_6 is a P.S. cell of dimension equal $\dim(Z_2) - 1$. Finally note that $N_6 = Z_5$ and $\dim(Z_2) - 1 = \dim(\Delta) - |J'| - 1$.

Now we consider Z_5 when $\Delta \not\subset \Delta_p$, and show that Z_5 is a P.S. cell of the desired dimension. Let N be as in Case 1. Note that N is a P.S. cell in ∂Z_2. Let $H^{(2)}$ also be as in Case 1. Set $\partial_+ Z_3 = |Star(N, H^{(2)})| \cap \partial Z_2$. Note that $\partial_+ Z_3$ is a regular neighborhood for the P.S. cell N in ∂Z_2. It follows that $\partial_+ Z_3$ is a P.S. cell of dimension equal $\dim(Z_2) - 1$. We have seen in Case 1 that Z_3 is a P.S. cell of dimension equal $\dim(Z_2)$. So we must have that $\partial_- Z_3$ (defined by $\partial_- Z_3 = \overline{\partial Z_3 - \partial_+ Z_3}$) is a P.S. cell of dimension equal $\dim(Z_2) - 1$ (see 10.5 and note $\partial_+ Z_3 \subset Z_3$). Finally note that

$$\partial_- Z_3 = Z_5,$$

so Z_5 is a P.S. cell of dimension equal $\dim(\Delta) - |J'| - 1$ as required.

Now a collapsing argument, similar to that used on pages 44-46 and pages 47–48, can be used to show that Z_6 is a P.S. cell of the desired dimension. The details are left to the reader.

An argument similar to the first two paragraphs of this case shows that $\left(\bigcap_{\Delta' \in J} \hat{Y}(\Delta')\right) \cap \left(\bigcap_{\Delta' \in J'} \partial \hat{Y}(\Delta')\right)$ is a P.S. cell of dimension $n - |J'|$.

This completes Step I in the proof of theorem 9.6.

STEP II. In this step we verify the following claim.

CLAIM 11.9. There is a P.S. embedding $r_j : V_j \to R^n$ which satisfies the following properties.

(a) Set $V_j = \cup Y'_\ell(\Delta)$, where the union runs over all $\Delta \in \cup_{x=0}^{n+m} K_{x,q}$, $1 \le \ell \le \phi(\Delta)$, $0 \le q \le j-1$. For each $\Delta \in \cup_{x=0}^{n+m} K_{x,j-1}$ and each $\ell \in \{1, 2, \ldots, \phi(\Delta)\}$ we must have that the image of $Y'_\ell(\Delta)$ under r_j is equal $Y_\ell(\Delta; 1)$. (See 11.5 for $Y'_\ell(\Delta)$ and $Y_\ell(\Delta; 1)$.)

(b) Set $Z_j = \cup Y'_\ell(\Delta)$, where the union runs over all $\{Y'_\ell(\Delta) : \Delta \in \cup_{x=0}^{n+m} K_{x,j-1},\ 1 \leq \ell \leq \phi(\Delta)\}$. Then $r_j|(V_j - Z_j) = $ identity.

(c) Suppose for some $Y'_\ell(\Delta)$ (as in (b)) we have for any other $Y'_{\ell'}(\Delta')$ (as in (b)) that either $Y'_{\ell'}(\Delta') = Y_{\ell'}(\Delta';1)$ or $Y'_{\ell'}(\Delta') \cap Y'_\ell(\Delta) = \emptyset$. Then $r_j|Y'_\ell(\Delta) = $ identity.

We begin by defining two P.S. homeomorphisms $s_j : R^n \to R^n$ and $t_j : R^n \to R^n$, and one P.S. embedding $u_j : U_j \to R^n$ (where U_j is defined in 11.12 below.) Then we set $r_j = u_j \circ t_j \circ s_j$.

Note that $Y_\ell(\Delta;1)$ is a union $Y'_\ell(\Delta) \cup \left(\cup_{(\Delta',\ell')\in J} Y'_{\ell'}(\Delta')\right)$ where J is a subset of $\{(\Delta',\ell') : \Delta' \in \cup_{x=0}^{n+m} K_{x,j},\ 1 \leq \ell' \leq \phi(\Delta')\}$ (see 11.5(d)). We let $\hat{Y}_\ell(\Delta;1)$ denote the corresponding union $Y'_\ell(\Delta) \cup \left(\cup_{(\Delta',\ell')\in J} \hat{Y}_{\ell'}(\Delta')\right)$. We construct now the P.S. homeomorphism $s_j : R^n \to R^n$ so as to satisfy the following properties.

<u>11.10</u> (a) The image of $Y'_\ell(\Delta)$ under $s_j : R^n \to R^n$ is equal $\hat{Y}_\ell(\Delta;1)$ for any $\Delta \in \cup_{x=0}^{n+m} K_{x,j-1}$ and $\ell \in \{1,2,\ldots,\phi(\Delta)\}$.

(b) $s_j|(V_j - Z_j) = $ identity (see 11.9 for V_j, Z_j).

(c) For any $Y'_\ell(\Delta)$ as in 11.9(c) we have that $s_j|Y'_\ell(\Delta) = $ identity.

In constructing $s_j : R^n \to R^n$ we shall make use of the ordering $X_1, X_2, \ldots, X_t$ of all the P.S. cells $\{\hat{Y}_{\ell'}(\Delta') : \Delta' \in \cup_{x=0}^{n+m} K_{x,j},\ 1 \leq \ell' \leq \phi(\Delta')\}$ given in 11.6 and lemma 10.7. For each $k \in \{1,2,\ldots,t\}$ we denote by ${}_k\hat{Y}_\ell(\Delta;1)$ the union $Y'_\ell(\Delta) \cup \left(\cup_{(\Delta',\ell')\in J_k} \hat{Y}_{\ell'}(\Delta')\right)$ where J_k is the subset of all $(\Delta',\ell') \in J$ such that $Y_{\ell'}(\Delta')$ is one of the $X_1, X_2, \ldots, X_k$. We proceed by induction over the index k; our induction assumption is that there is a P.S. homeomorphism ${}_ks_j : R^n \to R^n$ satisfying 11.10 (when in 11.10 we replace s_j, $\hat{Y}_\ell(\Delta;1)$ by ${}_ks_j$, ${}_k\hat{Y}_\ell(\Delta;1)$). Note that ${}_k\hat{Y}_\ell(\Delta;1) = {}_{k+1}\hat{Y}(\Delta;1)$ for all but at most one of the pairs (Δ,ℓ) (see 11.5(e)). If we have ${}_k\hat{Y}_\ell(\Delta;1) = {}_{k+1}\hat{Y}_\ell(\Delta;1)$ for all pairs (Δ,ℓ) then we set ${}_{k+1}s_j = {}_ks_j$. Otherwise there is a pair (Δ^*,ℓ^*) such that ${}_{k+1}\hat{Y}_{\ell^*}(\Delta^*) = {}_k\hat{Y}_{\ell^*}(\Delta^*) \cup X_{k+1}$. Note that 11.6(c)(d), our present induction hypothesis, and 11.5(a) allow us to apply lemma 10.7 to get a P.S. homeomorphism ${}_ks'_j : R^n \to R^n$ which maps each ${}_k\hat{Y}_\ell(\Delta;1)$ onto ${}_{k+1}\hat{Y}_\ell(\Delta;1)$. Now set ${}_{k+1}s_j = {}_{k+1}s'_j \circ {}_ks_j$. This completes the induction step in the construction of the $\{{}_ks_j : 1 \leq k \leq t\}$. Set $s_j = {}_ts_j$.

Now we construct the homeomorphism $t_j : R^n \to R^n$ which satisfies the following properties.

<u>11.11</u> (a) The image each $\Delta \in \cup_{x=0}^{n+m} K_{x,j}$ under t_j is equal to itself.

(b) For each $\Delta \in \cup_{x=0}^{n+m} K_{x,j}$ and each $1 \leq \ell \leq \phi(\Delta)$ we have that $t_j(\hat{Y}_\ell(\Delta)) = Y_\ell(\Delta)$, provided $Y_\ell(\Delta) \subset \mathrm{Int}(\cup_{x=0}^{n+m} |K_{x,j}|)$.

(c) $t_j(p) = p$ for each point p in $\overline{R^n - \cup_{x=0}^{n+m} |K_{x,j}|}$.

We first must discuss two partitions $D, \hat{D}$ of $\cup_{x=0}^{n+m} |K_{x,j}| = W$. Let V_1 be the smooth triangulation of W generated by all the $\{K_{x,j} : 0 \leq x \leq n+m\}$, and let V_2 be the partition of W consisting of all maximal non-empty intersections of sets of the form $\{W \cap Y_\ell(\Delta),\ \overline{W - Y_\ell(\Delta)},\ W \cap \partial Y_\ell(\Delta) : \Delta \in \cup_{x=0}^{n+m} K_{x,j},\ 1 \leq \ell \leq \phi(\Delta)\}$. Then D is the partition generated by V_1 and V_2. If in the description of V_2 we replace the $Y_\ell(\Delta)$ by the $\hat{Y}_\ell(\Delta)$ then we get $\hat{V}_2$; and $\hat{D}$ is the partition of W generated by V_1 and $\hat{V}_2$. Note that 11.4 and 11.6(a) make it possible to apply lemma 10.21 to conclude that for any $\Delta \in \cup_{x=0}^{n+m} K_{x,j}$ the partitions of Δ given by $\{\Delta \cap e : e \in V_2\}$ or by $\{\Delta \cap e : e \in \hat{V}_2\}$ are both P.S. regular cell structures for Δ. Thus each of D and $\hat{D}$ are P.S. regular cell structures for W. We also note that 11.4 and 11.6(a) assure us that the correspondence $\hat{Y}_\ell(\Delta) \to Y_\ell(\Delta)$ gives rise to a one-one correspondence $\hat{D} \to D$ satisfying the following:

(a) if $\hat{e} \to e$ then $\dim(e) = \dim(\hat{e})$;

(b) if $\hat{e}_1 \subset \hat{e}$ and $\hat{e}_1 \to e_1$, $\hat{e} \to e$ then $e_1 \subset e$.

Because the members of D and $\hat{D}$ are P.S. cells properties (a) and (b) are sufficient to assure the existence of a P.S. homeomorphism $t_j : \cup_{x=0}^{n+m} |K_{x,j}| \to \cup_{x=0}^{n+m} |K_{x,j}|$ which maps any $\hat{e} \in \hat{D}$ onto $e \in D$ if and only if $\hat{e} \to e$, and such that $t_j(\Delta) = \Delta$ for any $\Delta \in \cup_{x=0}^{n+m} K_{x,j}$. It is left as an exercise to check that t_j satisfies 11.11(a)(b). Note that t_j can be isotoped (near the boundary of $\cup_{x=0}^{n+m} |K_{x,j}|$) so that 11.11(a)(b) remain true and 11.11(c) becomes satisfied.

Now we construct the P.S. embedding $u_j : U_j \to R^n$. Let A denote all points of R^n contained in the interior of some minimal set of $\{Y_\ell(\Delta) : \Delta \in \cup_{x=0}^{n+m} K_{x,j},\ 1 \leq \ell \leq \phi(\Delta)\}$. Let B denote the partition of $Y = \cup Y_\ell(\Delta)$—where the union runs over all $\Delta \in \cup_{x=0}^{n+m} K_{x,j}$, $1 \leq \ell \leq \phi(\Delta)$—by sets of the form $\{Y - Y_\ell(\Delta),\ Y_\ell(\Delta) - \partial Y_\ell(\Delta), \partial Y_\ell(\Delta) : \Delta \in \cup_{x=0}^{n+m} K_{x,j},\ 1 \leq \ell \leq \phi(\Delta)\}$, and let $\bar{B}$ denote the collection of all closed sets $\bar{d}$ with $d \in B$ and $\bar{d} \subset A$. By replacing the balls $\{Y_\ell(\Delta)\}$ by the balls $\{Y'_\ell(\Delta)\}$ in the preceding construction we get the subsets A', Y' of R^n and the partitions $B', \bar{B}'$ (in place of the $A, Y, B, \bar{B}$). We require that $u_j : U_j \to R^n$ satisfy the following properties.

<u>11.12.</u> (a) $U_j = \cup_{\bar{d} \in \bar{B}} \bar{d}$.

(b) Note that the one to one correspondence $Y_\ell(\Delta) \to Y'_\ell(\Delta)$ induces a one to one correspondence $\bar{d} \to \bar{d}'$ for all $\bar{d} \in \bar{B}$ and $\bar{d}' \in \bar{B}'$. We require that $u_j(\bar{d}) = \bar{d}'$ for all $\bar{d} \in \bar{B}$.

(c) If for some $\bar{d} \in \bar{B}$ we have that $\bar{d}$ and all its faces are equal to $\bar{d}'$ and all its faces, then $u_j|\bar{d} =$ identity.

To construct such a $u_j : U_j \to R^n$ we simply note that 7.2, 8.5, 10.21 may be applied to conclude that both $B, \bar{B}'$ are regular cell complexes in R^n, and the correspondence $\bar{d} \to \bar{d}'$ of 11.12(b) is an isomorphism of cell complexes. So let $u_j : U_j \to R^n$ be a topological realization of this correspondence which satisfies 11.12(c).

Note that $r_j = u_j o t_j o s_j$ may be defined only on a subset $V_j' \subset V_j$ (see 11.9 for V_j). To extend r_j to all of V_j set $r_j|(V_j - V_j') =$ identity. It is left as an exercise to deduce from 11.10–11.12 and 11.5 that $r_j : V_j \to R^n$ is a well defined embedding satisfying 11.9(a)(b)(c).

This completes Step II.

STEP III. In this step we complete the proof of theorem 9.6.

We begin by constructing the P.S. embedding $h_{k,t} : T_{k,e} \to R^n$ of 9.6.

11.13. Set $h_{k,t} = r_t \circ r_{t-1} \circ r_{t-2} \circ \cdots \circ r_2 \circ r_1 \circ h_k$, where the r_j come from 11.9 and h_k comes from 8.5.

Now for each ${}_{k'}Y_{i',e}(f')$ as in 9.6 there is $\Delta \in \cup_{x=0}^{n+m} K_{x,0}$ and $\ell \in \{1, 2, \ldots, \phi(\Delta)\}$ such that $Y_\ell(\Delta) = {}_{k'}Y_{i',e}(f')$. Note that 11.13, 11.5, and the construction of the $K_{x,j}$ and $Y_{\ell'}(\Delta')$ guarantee us that the following property must hold.

11.14. ${}_{k'}Y_{i',e}(f';t) = \cup Y_{\tilde{\ell}}'(\tilde{\Delta})$, where the union runs over all $Y_{\tilde{\ell}}'(\tilde{\Delta})$ for which there exists a sequence of simplices $\Delta_1, \Delta_2, \ldots, \Delta_j$ with $j \leq t$ and a sequence of integers $\ell_1, \ell_2, \ldots, \ell_j$ which satisfy the following.

(a) $(\Delta_j, \ell_j) = (\tilde{\Delta}, \tilde{\ell})$; $\Delta_i \in \cup_{x=0}^{n+m} K_{x,i}$.

(b) If $j < t$ then $Y_{\tilde{\ell}}'(\tilde{\Delta}) \subset R^n - {}_rX_{u,e}$ for all $r > j+1$, and $Y_{\tilde{\ell}}(\tilde{\Delta}; 1) = Y_{\tilde{\ell}}'(\tilde{\Delta})$.

(c) For each $i \in \{1, 2, \ldots, j-1\}$ we set $Y_{\ell_i}(\Delta_i; 1)$ equal to the union $A_i \cup (\cup Y(A_i, \Delta'))$ of 11.5(d), where $A_i = Y_{\ell_i}'(\Delta_i)$. Then we must have that $Y_{\ell_{i+1}}(\Delta_{i+1})$ equals one of the $\{Y(A_i, \Delta') : \Delta' \in \cup_{x=0}^{n+m} K_{x,i+1}\}$.

(d) Set $Y_\ell(\Delta; 1)$ equal to the union $A_i \cup (\cup Y(A_i, \Delta')))$ of 11.5(d), where $A_i = Y_\ell'(\Delta)$. Then we must have that $Y_{\ell_1}(\Delta_1)$ is equal to one of the $\{Y(A_i, \Delta') : \Delta' \in \cup_{x=0}^{n+m} K_{x,1}\}$.

Now note that the conclusion of 9.6, that

$$h_{k,t}({}_{k'}Y_{i',e}(f')) = {}_{k'}Y_{i',e}(f';t),$$

follows from 11.14, 11.13, 11.9, and 8.5.

This completes the proof of theorem 9.6.

12. THE LIMIT THEOREM

In this section we prove the analogue of theorem 9.6 when $t = \infty$. Let ${}_{k'}Y_{i',e}(f;t)$ be as in 9.6. Then set

12.0 (a) ${}_{k'}Y_{i',e}(f;\infty) = \overline{\cup_{t\geq 1}\, {}_{k'}Y_{i',e}(f;t)}$
(b) If $k' \in I_e$ then set ${}_{k'}Y_{i'}(f;\infty) = {}_{k'}Y_{i',e}(f;\infty)$.

LIMIT THEOREM 12.1. *Let $e \in C$, $k \in I_e$, $S_{k,e}$ be as in 9.6. Let $S'_{k,e}$ denote all the ${}_{k'}Y_{i'}(f')$ in $S_{k,e}$ such that for each $x \in {}_{k'}Y_{i',e}(f')$ there is a minimum set in $\{{}_qY_{j,e}(g) : {}_qY_j(g) \in S_{k,e}\}$ which contains x on its interior. Set $T'_{k,e} = \cup_{k'}Y_{i',e}(f')$, where the union runs over all ${}_{k'}Y_{i'}(f')$ in $S'_{k,e}$. Then there is an embedding $h_{k,\infty} : T'_{k,e} :\longrightarrow R^n$ such that*

$$h_{k,\infty}({}_{k'}Y_{i',e}(f')) = {}_{k'}Y_{i',e}(f';\infty)$$

for each ${}_{k'}Y_{i'}(f')$ in $S'_{k,e}$.

Before beginning with the proof of theorem 12.1 we must state two lemmas. Let C_0 denote a regular cell complex in R^n, with $n = \dim(C_0)$.

DEFINITION 12.2. *A* **subdivision process** *for a finite regular cell complex C_0 consists of a sequence of finite regular cell complexes $C_1, C_2, C_3, \ldots$, satisfying the following properties.*

(a) $|C_{i+1}|$ is the underlying set of a subcomplex of C_i for all $i \geq 0$. If e is a cell of C_i such that $e \subset |C_{i+1}|$ then e is the underlying set of a subcomplex of C_{i+1}.

(b) For each $i \geq 0$, each cell $e \in C_i$ such that $e \subset |C_{i+1}|$, each subcomplex A of C_i which lies in ∂e, and each positive integer $j \in \{1, 2, \ldots, n+1\}$, we require that there is a regular neighborhood $N(A, e, j)$ for $|A|$ in e which meets the boundary ∂e regularly (see §10) and which satisfies property (c) below.

(c) There are positive integers $k_1 < k_2 < k_3 < \cdots < k_{n+2}$ such that

$$B(A, k_j) \cap e \subset N(A, e, j) \subset Int(B(A, k_{j+1})) \cap e$$

holds for all $j \in \{1, 2, \ldots, n+1\}$. (Here $n = \dim(C_0)$, and each $B(A, k)$ is a subcomplex of C_{i+1} defined as follows. Set $B(A, 1)$ equal to all cells $e \in C_{i+1}$ for which there is another $e' \in C_{i+1}$ with $e \cap e' \neq \emptyset$ and with $|A| \cap e' \neq \emptyset$. Given $B(A, k-1)$ define $B(A, k)$ to be all cells $e \in C_{i+1}$ such that there is another $e' \in C_{i+1}$ with $e \cap e' \neq \emptyset$ and $|B(A, k-1)| \cap e' \neq \emptyset$.)

(d) Each $d \in C_j$ is a P.L. cell in R^n for all $j \geq 0$. (Recall that $|C_0| \subset R^n$.)

DEFINITION 12.3. *C_i, $i \geq 1$, is a* **convergent subdivision process** *for C_0 if in addition to satisfying 12.2(a)-(d) the $\{C_i\}$ satisfy*

$$\lim_{i \to \infty} \beta_i = 0,$$

where β_i = maximum $\{\text{diameter}(e) : e \in C_i\}$ and all diameters are measured with a fixed metric $d(\ ,\ )$ on $|C_0|$.

Two subdivision processes for C_0, $\{C_i : i \geq 1\}$ and $\{C_i' : i \geq 1\}$, are said to be **isomorphic** if for each $i \geq 1$ there is a P.L. homeomorphism $g_i : R^n \to R^n$ satisfying the following:

(a) $g_i(e) = e$ for all $e \in C_0$ and all $i \geq 1$;

(b) For each $j \in \{1, 2, \ldots, i\}$ we have that g_i induces a cellular homeomorphism $C_j \to C_j'$.

(c) For any i, j with $i \geq j \geq 1$ and for any $e \in C_j$ we have that $g_i(e) = g_j(e)$.

LEMMA 12.4. *Any subdivision process for C_0, $\{C_i : i \geq 1\}$, is isomorphic to a convergent subdivision process for C_0, $\{C_i' : i \geq 1\}$.*

The second lemma of this section (see 12.6) states that a certain sequence of regular cell complexes $\{C_i : i \geq 1\}$ is a subdivision process for another regular cell complex C_0. We define now the $\{C_i : i \geq 0\}$ to which this next lemma applies.

Throughout the remainder of this section $e \in C$, $k \in I_e$ will be as in theorem 9.6. We consider the collection of all $\{{}_{k'}Y_{i',e}(f')\}$ as in 9.6. Let X_0 denote the subset of all $y \in R^n$ such that there is a minimum set in $\{{}_{k'}Y_{i',e}(f')\}$ that contains y on its interior. Let B_0 denote the partition of the union $Y_0 = \cup_{k'} Y_{i',e}(f')$ into maximal non-empty intersections of sets of the form $\{Y_0 - {}_{k'}Y_{i',e}(f'),\ {}_{k'}Y_{i',e}(f') - \partial_{k'}Y_{i',e}(f'),\ \partial_{k'}Y_{i',e}(f')\}$. Let B_0' denote the collection of all closed sets $\bar{d}$ with $d \in B_0$ and $\bar{d} \subset X_0$. Note it follows from 7.2 and 10.21 that B_0' is a regular cell complex.

We also must consider the smooth triangulations $\{K_{i,j} : 0 \leq i \leq n + m,\ j = 1, 2, \ldots\}$ of 11.2 and the smooth balls $\{Y_\ell'(\Delta) : 1 \leq \ell \leq \phi(\Delta),\ \Delta \in K_{i,j}\}$ of 11.5. Let B_j denote the partition of $Y_j = \cup Y_\ell'(\Delta)$—where this union runs over all $\Delta \in \cup_{i=0}^{n+m} K_{i,j}$ and over all $\ell \in \{1, 2, \ldots, \phi(\Delta)\}$—into maximal non-empty intersections of sets of the form $\{Y_j - Y_\ell'(\Delta),\ Y_\ell'(\Delta) - \partial Y_\ell'(\Delta),\ \partial Y_\ell'(\Delta) : \Delta \in \cup_{i=0}^{n+m} K_{i,j},\ 1 \leq \ell \leq \phi(\Delta)\}$. Let B_j' denote the collection of all closed sets $\bar{d}$ with $d \in B_j$ and $\bar{d} \subset Z_j$, where Z_j is the subset of all $y \in R^n$ such that there is a minimum set in $\{Y_\ell'(\Delta) : \Delta \in \cup_{i=0}^{n+m} K_{i,j},\ 1 \leq \ell \leq \phi(\Delta)\}$ that contains

y in its interior. Note it follows from 7.2 and 10.21 that B'_j is a regular cell complex.

12.5. (a) Set C_0 equal B'_0.

(b) For each $j > 0$ let C'_j denote the image of B'_j under the map $h_{k,j}^{-1}$ of 11.13. Then C_j is the maximal subcomplex of C'_j which satisfies the following: $|C_j|$ is the underlying set of a subcomplex of C_{j-1}.

LEMMA 12.6. *(a) The collection of regular complexes $\{C_i : i \geq 1\}$ of 12.5(b) is a subdivision process for the regular cell complex C_0 of 12.5(a).*

(b) Let $|\partial C_j|$ denote the topological boundary for $|C_j|$ in $|C_0|$ (note that ∂C_j is a subcomplex of C_j). Then $|\partial C_j| \cap |C_{j+1}| = \emptyset$ for all $j \geq 1$.

The proofs of lemmas 12.4 and 12.6 will be given at the end of this section. In the mean time we use lemmas 12.4, 12.6 to complete the proof of the limit theorem 12.1.

PROOF OF THEOREM 12.1. We have divided the proof into the following three steps.

STEP I. Set $h_{k,\infty} = \lim_{t\to\infty} h_{k,t}$, where the $h_{k,t}$ come from 11.13. In this step we prove that $h_{k,\infty}$ is a well defined continuous function.

Let the $r_j : V_j \to R^n$, $j = 1, 2, \ldots$, be as in 11.9. Then in view of 11.13 it will suffice to verify the following claim. Let $\lambda > 1$ and q be as in 5.1, 5.3.

CLAIM 12.7. $|r_j(x) - x| < \gamma\lambda^{-qj}$ holds for all $x \in V_j$, where γ is a positive number independent of x, j.

We begin by estimating the diameter of any of the balls $\{Y_\ell(\Delta) : \Delta \in \cup_{i=0}^{n+m} K_{i,j},\ 1 \leq \ell \leq \phi(\Delta)\}$ of 11.4. Let a, η be as in 5.1(b), 6.1(e). We denote by β the least upper bound of the diameters of all the balls $\{{}_kY_i(f) : i \in I_e,\ e \in C,\ f \in K_{u,e},\ 1 \leq k \leq y\}$ of 8.2. Note the following estimate can be deduced from 8.5(e), for q sufficiently large.

12.8 For any $\Delta \in \cup_{i=0}^{n+m} K_{i,j}$ and any $\ell \in \{1, 2, \ldots, \phi(\Delta)\}$ the diameter of $Y'_\ell(\Delta)$ is less than $2a^{-1}\beta\eta^2\lambda^{-qj}$.

Next we note that there are the following three possibilities for any $x \in V_j$ and any $j \in \{1, 2, \ldots\}$:

(a) $r_j(x) = x$;

(b) there is a ball $Y'_\ell(\Delta)$—with $\Delta \in \cup_{i=0}^{n+m} K_{i,j}$—such that $x \in Y'_\ell(\Delta)$ and $r_j(Y'_\ell(\Delta)) = Y'_\ell(\Delta)$;

(c) there are balls $Y'_\ell(\Delta)$, $Y'_{\ell'}(\Delta')$—with $\Delta \in \cup_{i=0}^{n+m} K_{i,j}$ and with $\Delta' \in \cup_{i=0}^{n+m} K_{i,j+1}$—such that $x \in Y'_\ell(\Delta)$, $r_j(x) \in Y'_{\ell'}(\Delta')$, and $Y'_\ell(\Delta) \cap Y'_{\ell'}(\Delta') \neq \emptyset$.

In each of these three cases it follows from 12.8 that

$$|r_j(x) - x| < 2a^{-1}\beta\eta^2\lambda^{-qj} + 2a^{-1}\beta\eta^2\lambda^{-q(j+1)}.$$

This inequality yields 12.7 for

12.9 $\gamma = 4a^{-1}\beta\eta^2$.

This completes step I.

STEP II. In this step we complete the proof of 12.1 under the following assumption.

ASSUMPTION 12.10. The subdivision process $\{C_j : j \geq 1\}$ for C_0 given in lemma 12.6 is a convergent subdivision process.

For any $j \geq 0$ and any cells $d_1, d_2 \in C_j$ we will say that d_1 and d_2 are **ball-disjoint** if there are balls $Y_{\ell_1}(\Delta_1)$, $Y_{\ell_2}(\Delta_2)$ in $\{Y_\ell(\Delta) : \Delta \in \cup_{i=0}^{n+m} K_{i,j},\ 1 \leq \ell \leq \psi(\Delta)\}$ such that $h_{k,j}(d_i) \subset Y_{\ell_i}(\Delta_i)$, $i = 1, 2$, and $Y_{\ell_1}(\Delta_1) \cap Y_{\ell_2}(\Delta_2) = \emptyset$.

We define a positive number β' to be the greatest lower bound of all the numbers $\varepsilon > 0$ which satisfy the following property: For any $e \in C$ consider the collection of balls $\{{}_{k'}Y_{i',e}(f)\}$ in R^n of 8.5(b); then if two such balls don't intersect they must be at least a distance ε apart. The following properties of β' are deduced from 7.2, 8.2, 12.5(a).

12.11 (a) There is a positive lower bound ω for the ratios β/β', β'/β, where β comes from 12.9. Moreover ω depends only on μ of 7.2, and is therefore independent of q.

(b) If d_1, d_2 are ball-disjoint closed cells of C_0 then the distance from d_1 to d_2 is at least β'.

Note it follows from 11.13, 12.7, 8.5, 12.9 that $h_{k,\infty} : T'_{k,e} \to R^n$ (defined in step I) satisfies the following property.

12.12 $|h_{k,\infty}(x) - x| < 5\phi a^{-1}\beta\eta^2 \left(\frac{1}{1-\lambda^{-q}}\right)\lambda^{-q}$, for all $x \in T'_{k,e}$, where ϕ comes from 8.5(e).

Now we deduce from 12.12, 8.5, 12.11(a) that for q sufficiently large we must have the following.

12.13 $|h_{k,\infty}(x) - x| << \beta'$, for all $x \in T'_{k,e}$.

Combining 12.11(b) and 12.13 we have the following.

12.14 If d_1, d_2 are ball-disjoint closed cells in C_0 and $x_1 \in d_1$, $x_2 \in d_2$ then $h_{k,\infty}(x_1) \neq h_{k,\infty}(x_2)$.

There is the following extension of property 12.14.

12.15 (a) Suppose d_1, d_2 are ball-disjoint closed cells in C_j and $x_1 \in d_1$, $x_2 \in d_2$. Then $h_{k,\infty}(x_1) \neq h_{k,\infty}(x_2)$.

(b) Suppose $x_1 \in |C_0| - \cap_{k>0}|C_k|$ and $x_2 \in \cap_{k\geq 0}|C_k|$. Then $h_{k,\infty}(x_1) \neq h_{k,\infty}(x_2)$.

To verify 12.15(a) we note that since d_1, d_2 are ball-disjoint there must be balls $Y_{\ell_1}(\Delta_1)$, $Y_{\ell_2}(\Delta_2)$ with $\Delta_1, \Delta_2 \in \cup_{i=0}^{n+m} K_{i,j}$ such that $d_1 \subset h_{k,j}^{-1}(Y_{\ell_1}(\Delta_1))$, $d_2 \subset h_{k,j}^{-1}(Y_{\ell_2}(\Delta_2))$, and $Y_{\ell_1}(\Delta_1) \cap Y_{\ell_2}(\Delta_2) = \emptyset$. On the other hand if $h_{k,\infty}(x_1) = h_{k,\infty}(x_2)$ in 12.15(a) then we must have that $Y_{\ell_1}(\Delta_1;\infty) \cap Y_{\ell_2}(\Delta_2;\infty) \neq \emptyset$, where $Y_{\ell_1}(\Delta_1;\infty)$ and $Y_{\ell_2}(\Delta_2;\infty)$ are the images of $Y_{\ell_1}(\Delta_1)$ and $Y_{\ell_2}(\Delta_2)$ under the continuous map

$$\lim_{y\to\infty}(r_{j+y} \circ r_{j+y-1} \circ r_{j+y-2} \circ \cdots \circ r_{j+1}).$$

(Note that the above limit is a well defined continuous function because of 12.7.) We denote by X_1 and X_2 the images of $g_e(Y_{\ell_1}(\Delta_1;\infty) \times R^n)$ and $g_e(Y_{\ell_2}(\Delta_2;\infty) \times R^n)$ under the diffeomorphism $F^{jq}: M \to M$. Note there is another pair (e', k')—with $e' \in C$ and $k' \in I_{e'}$—such that the images of $X_1 \cap g_{e'}(R^n \times R^m)$ and $X_2 \cap g_{e'}(R^n \times R^m)$ (denoted by X_1' and X_2'), under the composite map

$$g_{e'}(R^n \times R^m) \xrightarrow{g_{e'}^{-1}} R^n \times R^m \xrightarrow{\text{proj.}} R^n,$$

satisfy the following properties.

12.16 Let C_0' denote the cell structure in R^n which is constructed as in 12.5(a) for the pair (e', k') in place of the pair (e, k).

(a) There are subcomplexes Y_1, Y_2 of C_0' such that for any $d_1 \in Y_1$, $d_2 \in Y_2$ we have that d_1, d_2 are ball-disjoint.

(b) Let $h_{k',\infty}: R^n \to R^n$ be the homeomorphism constructed as in step I for the pair (e', k') in place of the pair (e, k). Then $h_{k',\infty}(|Y_1|) = X_1'$ and $h_{k',\infty}(|Y_2|) = X_2'$.

(c) $X_1' \cap X_2' \neq \emptyset$.

Note that 12.16 contradicts 12.14 (when in 12.14 we replace $C_0, h_{k,\infty}$ by $C_0', h_{k',\infty}$). This contradiction can only be avoided if $X_1 \cap X_2 = \emptyset$, which would imply that $h_{k,\infty}(x_1) \neq h_{k,\infty}(x_2)$ as required in 12.15(a).

To verify 12.15(b) we reduce the problem to the situation in 12.15(a). Without loss of generality we may assume that $x_1 \in |C_j|$, but $x_1 \notin |C_{j+1}|$ for some $j \geq 0$. Choose cells $d_1, d_2 \in C_j$ such that $x_1 \in d_1$ and $x_2 \in d_2$. Now if the cells d_1, d_2 are ball-disjoint in C_j then 12.15(a) applies to verify 12.15(b). On the other hand if d_1, d_2 are not ball-disjoint in C_j then it follows from 11.0, and the fact that $x_1 \in |C_j|$ but $x_1 \notin |C_{j+1}|$, that $x_2 \notin |C_{j+2}|$. This is a contradiction of the hypothesis of 12.15(b).

We can now complete step II. We do this by showing that $h_{k,\infty}$: $\text{Int}(|C_0|) \to R^n$ is one-one. Note it follows from 11.0 and from the definition of C_0 that $T'_{k,e} \subset |C_0|$. So $h_{k,\infty} : T'_{k,e} \to R^n$ will have been shown to be one-one as is required by the conclusion of step II.

There are the following three cases to consider.

CASE I. In this case we suppose that $x_1, x_2 \in (\cap_{j=0}^{\infty}|C_j|)$ and that $x_1 \neq x_2$. We show that $h_{k,\infty}(x_1) \neq h_{k,\infty}(x_2)$. Note that it will suffice (by 12.15(a)) to show that for some $j \geq 0$ there are ball-disjoint cells $d_1, d_2 \in C_j$ with $x_1 \in d_1$ and $x_2 \in d_2$.

For each $j \geq 0$ and any ball $Y_k(\Delta)$, with $\Delta \in \cup_{i=0}^{n+m} K_{i,j}$ and with $k \in \{1, 2, \ldots, \psi(\Delta)\}$, let $Y_k^\star(\Delta)$ denote the maximal subcomplex of C_j such that for any $d \in Y_k^\star(\Delta)$ we have that $h_{k,j}(d) \subset Y_k(\Delta)$. Choose $d_1, d_2 \in C_j$, and balls $Y_{k_1}(\Delta_1), Y_{k_2}(\Delta_2)$ with $\Delta_1, \Delta_2 \in \cup_{i=0}^{n+m} K_{i,j}$ and $k_1 \in \{1, 2, \ldots, \psi(\Delta_1)\}$, $k_2 \in \{1, 2, \ldots, \psi(\Delta_2)\}$, such that $x_1 \in d_1, x_2 \in d_2$, $d_1 \in Y_{k_1}^\star(\Delta_1)$, $d_2 \in Y_{k_2}^\star(\Delta_2)$. Note that there is an upper bound N to the number of cells in $Y_{k_1}^\star(\Delta_1)$ and in $Y_{k_2}^\star(\Delta_2)$, where N is independent of $j, k_1, k_2, \Delta_1, \Delta_2$. So it follows from 12.10 that for sufficiently large j any connected component of $|Y_{k_1}^\star(\Delta_1)|$ or of $|Y_{k_2}^\star(\Delta_2)|$ has diameter less than $\frac{1}{4}d(x_1, x_2)$. Hence if each of $h_{k,j} : |Y_{k_1}^\star(\Delta_1)| \to Y_{k_1}(\Delta_1)$ and $h_{k,j} : |Y_{k_2}^\star(\Delta_2)| \to Y_{k_2}(\Delta_2)$ are homeomorphisms then the d_1, d_2 must be ball-disjoint cells in C_j, which would complete the argument of case I. On the other hand if (for example) we have that $h_{k,j} : |Y_{k_1}^\star(\Delta_1)| \to Y_{k_1}(\Delta_1)$ is not a homeomorphism (for arbitrarily large j) then one can deduce from 11.0 that $x_1 \notin \cap_{j=0}^{\infty}|C_j|$, which contradicts the hypothesis of case I. These remaining details are left to the reader to verify (see 11.0).

CASE 2. Suppose $x_1 \in \cap_{j=1}^{\infty}|C_j|$, $x_2 \in |C_0| - \cap_{j=1}^{\infty}|C_j|$. Then by 12.15(b) we must have $h_{k,\infty}(x_1) \neq h_{k,\infty}(x_2)$.

CASE 3. Suppose $x_1, x_2 \in |C_0| - \cap_{j=1}^{\infty}|C_j|$, but $x_1 \neq x_2$. Note it follows from 11.0, 11.9(b), 11.13, 12.5, 12.6 that for any $j \in \{1, 2, \ldots\}$ we must have that

$$h_{k,\infty}|(|C_0| - |C_{j-1}|) = h_{k,j}|(|C_0| - |C_{j-1}|).$$

Choose j sufficiently large that $x_1, x_2 \in |C_0| - |C_{j-1}|$. Then by the above equality we have that $h_{k,\infty}(x_1) = h_{k,j}(x_1)$ and $h_{k,\infty}(x_2) = h_{k,j}(x_2)$, from which we conclude $h_{k,\infty}(x_1) \neq h_{k,\infty}(x_2)$ because $h_{k,j}$ is a homeomorphism.

Cases 1,2,3 above complete the proof (when 12.10 holds) of that part of 12.1 which asserts that $h_{k,\infty} : T'_{k,e} \to R^n$ is an embedding. The other property asserted by 12.1 for $h_{k,\infty}$ (that $h_{k,\infty}({}_{k'}Y_{i',e}(f)) = {}_{k'}Y_{i',e}(f,\infty)$) is a consequence of the definitions of $h_{k,\infty}$, ${}_{k'}Y_{i',e}(f;\infty)$, and 9.6.

This completes step II.

STEP III. In this step we complete the proof of 12.1. If $\{C_j : j \geq 1\}$ is a convergent subdivision process we are done (step II). Otherwise use lemma 12.4 to choose another subdivision process $\{C'_j : j \geq 1\}$ for C_0 satisfying the following properties.

12.17. (a) $\{C'_j : j \geq 1\}$ is a convergent subdivision process for C_0.
(b) $\{C'_j : j \geq 1\}$ is isomorphic to $\{C_j : j \geq 1\}$ via an isomorphism consisting of homeomorphisms $g_j : R^n \to R^n$, $j = 1, 2, \ldots$, as preceding 12.4.

Define embeddings $h'_{k,j} : \tilde{T}_{k,e} \to R^n$ to be the composite of maps

$$R^n \xrightarrow{g_j^{-1}} R^n \supset T_{k,e} \xrightarrow{h_{k,j}} R^n.$$

We leave as an exercise the verification of the following properties for the $\{h'_{k,j} : j \geq 1\}$ (use 12.4 and the results in steps I and II above).

12.18. (a) If in 12.5 we replace the $\{h_{k,j} : j \geq 1\}$ by the $\{h'_{k,j} : j \geq 1\}$, then the subdivision process for C_0 which is obtained is equal to the $\{C'_j : j \geq 1\}$ of 12.17.
(b) $h'_{k,\infty} = \lim_{t\to\infty} h'_{k,t}$ is a well defined continuous map.
(c) 12.15(a)(b) are satisfied for $h'_{k,\infty}$ and the $\{C'_j : j \geq 1\}$.

To complete the proof of theorem 12.1 it will suffice to show that $h'_{k,\infty} : |B'_0| \to R^n$ is one-one. Using 12.17(a), 12.18(c) we can argue exactly as in the three cases at the end of step II to show that $h'_{k,\infty}$ is one-one. Now set $h_{k,\infty} = h'_{k,\infty}$ in 12.1.

This completes the proof of theorem 12.1.

PROOF OF LEMMA 12.4. The proof is divided into the following five steps.

STEP I. In steps I,II,III e will denote a fixed cell in C_0. For each positive integer j we denote by $C_{0,j}$ the unique regular cell complex which satisfies the following properties.
(a) $C_{0,j}$ is a subdivision of C_0.

(b) For each $k \in \{0,1,2,\ldots,j\}$ C_k is the underlying set of a subcomplex of $C_{0,j}$, denoted by $C_{0,k,j}$. Moreover $C_{0,k,j}$ is a subdivision of C_k. C_j is a subcomplex of $C_{0,j}$.

(c) $C_{0,j}$ is the complex with the least number of cells which satisfies (a) and (b).

DEFINITION 12.19. *A regular neighborhood N for ∂e in e is said to be* **refined by** $C_{0,j}$ *if the following properties hold.*

(a) For each $d \in C_{0,j}$ with $d \subset e$ we have that $N \cap d$ is a regular neighborhood for $d \cap \partial e$ in d.

(b) Moreover the regular neighborhood $N \cap d$ meets the boundary ∂d regularly (see §10).

CLAIM 12.20. For any positive integers j, x there is a sequence $N_1, N_2, \ldots, N_x$ of regular neighborhoods for ∂e in e which are refined by $C_{0,j}$ and such that $N_i \subset \mathrm{Int}(N_{i+1})$ holds for all $i \in \{1,2,\ldots,x-1\}$. Moreover if $N_1', N_2', \ldots, N_x'$ is another such sequence of regular neighborhoods for ∂e in e then there is an isotopy $\phi_{e,t} : e \to e$, $t \in [0,1]$, satisfying the following properties.

(a) $\phi_{e,0} =$ identity; $\phi_{e,t}|\partial e =$ identity for all $t \in [0,1]$.

(b) $\phi_{e,t}(d) = d$ holds for all $t \in [0,1]$ and all cells $d \in C_{0,j}$.

(c) $\phi_{e,1}(N_i) = N_i'$ holds for all $i \in \{1,2,\ldots,x\}$.

The verification of claim 12.20 is the purpose of step I.

First we will construct N_1 of 12.20. We carry out this construction by induction over the dimension of the cells of $C_{0,j}$ which lie in e. Suppose that for all such cells $d \in C_{0,j}$ with $\dim(d) \leq r$ the set $N_1 \cap d$ has already been constructed so that $N_1 \cap d$ is a regular neighborhood for $d \cap \partial e$ in d which meets the boundary ∂d regularly. Let $N_{1,r}$ denote the union $\cup(N_1 \cap d)$ taken over all $d \in C_{0,j}$ with $\dim(d) \leq r$ and $d \subset e$. Note that for any cell $d' \in C_{0,j}$ with $d' \subset e$ and $\dim(d') = r+1$ we have that $N_{1,r} \cap d'$ is a regular neighborhood for $\partial d' \cap \partial e$ in $\partial d'$. Thus we can extend $N_{1,r} \cap d'$ to a set $N_1 \cap d'$ such that $N_1 \cap d'$ is a regular neighborhood for $d' \cap \partial e$ in d' and such that $(N_1 \cap d') \cap \partial d' = N_{1,r} \cap d'$. This completes the induction step in the construction of $N_1 = N_{1,n}$.

To get N_2 we proceed by induction as in the preceding paragraph. However we now require that $N_2 \cap d$ is a regular neighborhood for $N_1 \cap d$ in d which meets the boundary ∂d regularly. The same construction gives $N_3, N_4, \ldots, N_x$.

To get the isotopy $\phi_{e,t} : e \to e$ of 12.20 we again proceed by induction over the dimension of the cells of $C_{0,j}$ which lie in e. Suppose that $\phi_{e,t} : d \to d$, $t \in [0,1]$, has been constructed for all cells $d \in C_{0,j}$ with $d \subset e$ and $\dim(d) \leq r$, so that 12.20(a)(b)(c) hold wherever $\phi_{e,t}$ has been defined (namely on $e \cap |C^r_{0,j}|$). Let d' be an $(r+1)$-cell in $C_{0,j}$ with $d' \subset e$. Use the isotopy extension theorem (see theorem 6.12 in [12]) to extend $\phi_{e,t} : \partial d' \to \partial d'$ to a P.L. isotopy $f_{d',t} : d' \to d'$, $t \in [0,1]$. Note that both $N'_1 \cap d'$ and $f_{d',1}(N_1 \cap d')$ are regular neighborhoods for $d' \cap \partial e$ in d' which meet the boundary $\partial d'$ regularly in the same subset. Thus by the uniqueness of regular neighborhoods (see theorem 10.1, 10.2) there must be another P.L. isotopy $g^1_{d',t} : d' \to d'$, $t \in [0,1]$, such that $g^1_{d',t}|\partial d' =$ identity for all $t \in [0,1]$ and such that $g^1_{d',1}(f_{d',1}(N_1 \cap d'))$ equals $N'_1 \cap d'$. By applying theorem 10.1, 10.2 over and over again we get for each $i \in \{1, 2, \ldots, x\}$ a P.L. isotopy $g^i_{d',t} : d' \to d'$, $t \in [0,1]$, such that $g^i_{d',t} : (N'_{i-1} \cap d') \cup \partial d' \to (N'_{i-1} \cap d') \cup \partial d'$ is the identity map and such that $g^i_{d',1} \circ g^{i-1}_{d',1} \circ \cdots \circ g^1_{d',1} \circ f_{d',1}(N_1 \cap d')$ is equal to $N'_i \cap d'$. Now by composing all the isotopies $f_{d',t}, g^1_{d',t}, g^2_{d',t}, \ldots, g^x_{d',t}$ and reparametrizing all of these isotopies (as well as reparametrizing $\phi_{e,t}|\,|C^r_{0,j}| \cap e$) we get the desired extension of $\phi_{e,t}|\partial d'$ to $\phi_{e,t}|d'$. This completes the induction step in the construction of $\phi_{e,t} : e \to e$.

This completes step I in the proof of lemma 12.4.

STEP II. Let j, x and $N_1, N_2, \ldots, N_x$ be as in claim 12.20. In this step we show that the $N_1, N_2, \ldots, N_x$ can be chosen to satisfy the following additional property.

12.21. For each integer $q > j$ let e_q denote the subcomplex of C_q consisting of the union of all closed cells $d \in C_q$ such that $d \subset e$ and $d \cap \partial e \neq \emptyset$. Then the $N_1, N_2, \ldots, N_x$ must satisfy the following properties.

(a) N_i is a regular neighborhood for ∂e in e which is refined by $C_{0,j+x-i}$. (Note that this is stronger than the requirement of 12.20 that N_i be refined by $C_{0,j}$.)

(b) $e_{j+x-i+1} \subset \text{Int}(N_i)$.

We will construct N_i for any $i \in \{1, 2, \ldots, x\}$ by using the procedure of step I that yielded N_1. Thus we proceed by induction over the dimension of the cells of $C_{0,j+x-i}$ which lie in e. Suppose that $N_i \cap d$ has been constructed for any cell $d \in C_{0,j+x-i}$ with $d \subset e$ and with $\dim(d) \leq r$. We also make the assumption (and this is where our present argument differs somewhat from step I) that for all $d \in C_{j+x-i}$ with $d \subset e$ and $\dim(d) = s \leq r$ we must have that

12.22. $d \cap B(A, k_{n-s+1}) \subset N_i \cap d \subset d \cap B(A, k_{n+2})$,

where $A = d \cap \partial e$, and $B(A, k_{n-s+1})$ comes from 12.2(c) where in 12.2(c) we replace $i+1$ by $j+x-i+1$. Let d' denote any cell in $C_{0,j+x-i}$ having $\dim(d') = r+1$ and having $d' \subset e$. Note that 12.2(c) allows us to extend the set $N_i \cap \partial d'$ (which is a regular neighborhood for $\partial d' \cap \partial e$ in $\partial d'$) to a regular neighborhood $N_i \cap d'$ for $d' \cap \partial e$ in d' such that if $d' \in C_{j+x-i}$ then we must have that

12.23. $d' \cap B(A', k_{n-s}) \subset N_i \cap d' \subset d \cap B(A, k_{n+2})$, where $A' = d' \cap \partial e$.

This completes the induction step in the construction of any N_i. Note it follows from 12.2(c), 12.22, 12.23 that for each fixed i the N_i just constructed satisfies 12.21(a)(b). There is also no loss of generality in supposing that $N_i \subset \mathrm{Int}(N_{i+1})$ holds for each i, as is required by 12.20 (the details are here left to the reader).

This completes step II.

STEP III. For $e \in C_0$ as in steps I, II set $\dim(e) = m$. Let X denote a linear m-simplex in R^n having the points $v_1, v_2, \ldots, v_{m+1}$ for vertices. Subdivide X by adding the barycenter of X—denoted by b—to the existing set of vertices of X, and denote by Y the resulting linear triangulation in R^n having $v_1, v_2, \ldots, v_{m+1}, b$ as its vertices. Let

12.24. $h : Y \to [0, 1]$

denote the simplicial map which maps the $v_1, \ldots, v_{m+1}$ to 0 and maps the vertex b to 1.

Choose a P.L. homeomorphism $g : e \to Y$. The verification of the following claim—the statement of which is the purpose of step III—is left to the reader.

CLAIM 12.25. Given any integer $j > 0$ there is a number $\varepsilon \in (0, 1)$ which depends only on $g : e \to Y$ and on j. For any $\delta \in (0, \varepsilon)$ the set $N = (h \circ g)^{-1}([0, \delta])$ is a regular neighborhood for ∂e in e which is refined by $C_{0,j}$.

This completes step III.

STEP IV. Let $e \in C_0$ be as in steps I, II, III, and for each $j \geq 0$ let $e(j)$ denote the maximal subcomplex of C_j which is contained in ∂e. In this step we verify the following claim.

CLAIM 12.26. Given any $\varepsilon > 0$ there is $\delta \in (0, \varepsilon)$ such that the following hold. Suppose that for some $j > 0$ the diameter of every cell $d \in e(j)$ is less than δ. Then there is an integer $j' > j$ and a P.L. isotopy $\psi_t : e \to e$, $t \in [0, 1]$, which satisfies the following properties.

(a) $\psi_0 =$ identity; $\psi_t | \partial e =$ identity for all $t \in [0, 1]$.

(b) For each cell $d \in C_{j'}$ with $d \subset e$ we have that the diameter of $\psi_1(d)$ is less than ε.

Choose ε_1 sufficiently small so that for any $\delta_1 \in (0, \varepsilon_1)$ we have that $(h \circ g)^{-1}([0, \delta_1])$ is a regular neighborhood for ∂e in e which is refined by C_j, where j comes from 12.26 (see 12.25). For any positive integer x define $N'_1, N'_2, \ldots, N'_x$ by

$$N'_i = (h \circ g)^{-1}([0, \varepsilon_1(i/x + 1)]).$$

Let $N_1, N_2, \ldots, N_x$ be as in 12.20 and 12.21, and let $\phi_{e,t} : e \to e,\ t \in [0,1]$, be as in 12.20.

Choose a P.L. isotopy $r_t : X \to X,\ t \in [0,1]$, satisfying the following properties. (Recall that $X = |Y|$.)

12.27. (a) $r_0 =$ identity; $r_t|\partial X =$ identity for all $t \in [0,1]$.

(b) $r_1(h^{-1}([0, i\varepsilon_1/(x+1)])) = h^{-1}([0, i/(x+1)])$ for all $i \in [1, 2, \ldots, x]$.

(c) $r_t(L) = L$ for all $t \in [0,1]$ and all rays L which begin at a point of ∂X and end at the barycenter of X.

Now define $\psi_t : e \to e,\ t \in [0,1]$, of 12.26 as follows.

12.28.

$$\psi_t(p) = \begin{cases} \phi_{e,2t}(p), & \text{if } t \in [0, 1/2], \\ g^{-1} \circ r_{2(t-1/2)} \circ g \circ \phi_{e,1}(p), & \text{if } t \in [1/2, 1]. \end{cases}$$

Note it follows from 12.27, 12.25, 12.21, 12.20 that the isotopy defined in 12.28 satisfies the conclusions of 12.26, provided ε_1 in 12.27 is chosen sufficiently small and the integer x is chosen sufficiently large. Note that j' may be chosen to equal $j + x + 4$ in 12.26.

This completes step IV.

STEP V. In this step we complete the proof of lemma 12.4.

The construction of the $g_i : R^n \to R^n$ and of the C'_i of 12.4 proceeds by induction over the iteger $r \geq 0$ in the following induction hypothesis.

12.29(r). For each $i \in \{1, 2, 3, \ldots, r\}$ there is an integer $i' > i$ and a P.L. homeomorphism $h_i : R^n \to R^n$ which satisfy the following properties.

(a) For any $i, j \in \{1, 2, \ldots, r\}$ with $i < j$ and for any $d \in C_{i'}$ we have that $i' < j'$ and $h_i(d) = h_j(d)$.

(b) For any $i \in \{1, 2, \ldots, r\}$ and any $d \in C_{i'}$ we have that the diameter of $h_i(d)$ is less than $1/i$.

We will now carry out the induction step 12.29(r) $\Rightarrow$ 12.29($r+1$). The argument which verifies this induction step is itself an induction argument whose induction step consists of an application of 12.26. Here is a brief description of this second induction argument. We will first construct $h_{r+1}|d$ for each cell in $C_{0,r'}$ by induction over the dimension of the cells d. Here is our second induction hypothesis.

12.30(s). $h_{r+1}|d$ has been constructed for all $d \in C_{0,r'}$ such that $\dim(d)$
$\leq s$, so that the following properties hold.

(a) For some very large integer $I_{r,s}$, and for each $d' \in C_{I_{r,s}}$ with $d' \subset d$, we have that $h_{r+1}(d')$ has very small diameter.

(b) $h_{r+1}(d) = h_r(d)$.

(c) There is an isotopy $H_t^s : h_r(|C_{0,r'}|) \to h_r(|C_{0,r'}|)$, $t \in [0,1]$, satisfying: $H_0^s =$ identity; $H_t^s \circ h_r(d) = h_r(d)$ for all $t \in [0,1]$ and all $d \in C_{0,r'}^s$; $H_1^s \circ h_r|d = h_{r+1}|d$ for all $d \in C_{0,r'}^s$.

To carry out the induction step 12.30(s) $\Rightarrow$ 12.30($s+1$) apply 12.26 to each cell $e \in C_{0,r'}$. In more detail, we set $C'_{r'+i}$ equal the image of $C_{r'+i}$ under the map $H_1^s \circ h_r : |C_{0,r'}| \to |C_{0,r'}|$, for $i \geq 0$.

We note that $C'_{r'+1}, C'_{r'+2}, C'_{r'+3}, \ldots$ is a subdivision process for $C'_{r'}$. Thus we can replace in 12.26 the $C_0, C_1, C_2, C_3, C_4, \ldots$ by the subdivision process $C'_{r'}, C'_{r'+1}, C'_{r'+2}, C'_{r'+3}, C'_{r'+4}, \ldots$ and apply 12.26 to any $e \in C_{r'}$ to get an isotopy $\psi_{e,t} : e' \to e'$, $t \in [0,1]$, satisfying the following (here $e' = H_1^s \circ h_r(e)$).

12.31. (a) $\psi_{e,0} =$ identity; $\psi_{e,t}|\partial e' =$ identity for all $t \in [0,1]$.

(b) For some very large integer $I_{r,s+1}$ and for any cell $d' \in C_{I_{r,s+1}}$ with $d' \subset e$ we have that $\psi_{e,1} \circ H_1^s \circ h_r(d')$ has very small diameter.

If $e \in C_{0,r'}$ but $e \notin C_{r'}$ then we define $\psi_{e,t} : e' \to e'$, $t \in [0,1]$—where $e' = H_1^s \circ h_r(e)$—by

12.31. (c) $\psi_{e,0} =$ identity for all $t \in [0,1]$.

Now we define the isotopy $H_t^{s+1} : h_r(|C_{0,r'}|) \to h_r(|C_{0,r'}|)$ of 12.30(s+1) as follows. Let $\psi_t : h_r(|C_{0,r'}^{s+1}|) \to h_r(|C_{0,r'}^{s+1}|)$, $t \in [0,1]$, denote the union of the $\psi_{e,t}$ in 12.31, where the union runs over all $e \in C_{0,r'}$ with $\dim(e) = r+1$. Use the isotopy extension theorem to extend $\psi_t : h_r(|C_{0,r'}^{s+1}|) \to h_r(|C_{0,r'}^{s+1}|)$ to an isotopy $\psi_t : h_r(|C_{0,r'}|) \to h_r(|C_{0,r'}|)$ such that $\psi_t(d') = d'$ for all $t \in [0,1]$ and all $d \in C_{0,r'}$, where $d' = H_1^s \circ h_r(d)$. Now define $H_t^{s+1} : h_r(|C_{0,r'}|) \to h_r(|C_{0,r'}|)$ to be

12.32.

$$H_t^{s+1}(p) = \begin{cases} H_{2t}^s(p), & \text{if } t \in \left[0, \frac{1}{2}\right], \\ \psi_{2(t-1/2)} \circ H_1^s(p), & \text{if } t \in \left[\frac{1}{2}, 1\right]. \end{cases}$$

We can now define $h_{r+1}|d$, for $d \in C_{0,r'}$ with $\dim(d) = s+1$, to be equal $H_1^{s+1} \circ h_r|d$. Note that it follows from 12.30(s), 12.31, and 12.32 that h_{r+1} and H_t^{s+1} satisfies 12.30(s+1).

This completes the induction step 12.30(s) $\Rightarrow$ 12.30($s+1$).

We now return to the verification of the induction step 12.29(r) $\Rightarrow$ 12.29($r+1$).

Let $H_t^n : h_r(|C_{0,r'}|) \to h_r(|C_{0,r'}|)$, $t \in [0,1]$, be as in 12.30(n), where $n = \dim(C_{0,r'})$. Let $C'_{0,r'}$ denote the image of $C_{0,r'}$ under the map $h_r : R^n \to R^n$. There is no loss of generality in assuming that the cell structure $C'_{0,r'}$ extends to a P.L. cell structure C for all of R^n. (To verify this last claim the reader should review the definition of the C_j and the $C_{0,j}$.) Thus we may use the isotopy extension theorem to extend $H_t^n : |C'_{0,r'}| \to |C'_{0,r'}|$ to an isotopy $H_t : R^n \to R^n$ which satisfies $H_t(d) = d$ for all $t \in [0,1]$ and all $d \in C$. Now define $h_{r+1} : R^n \to R^n$ in 12.29($r+1$) to equal the composition $H_1 \circ h_r : R^n \to R^n$. It follows from 12.29($r$) and 12.30($n$) that h_{r+1} satisfies 12.29($r+1$).

This completes the induction step 12.29(r) $\Rightarrow$ 12.29($r+1$). Note that we may take $(r+1)'$ equal to $I_{r,n}$ in 12.29(r+1).

We can now complete the proof of lemma 12.4. Set:

12.33.

$$g_i = h_i;$$
$$C'_i = \text{the image of } C_i \text{ under the map } g_i : R^n \to R^n.$$

Note that it follows from 12.29(i), for $i \geq 1$, that the g_i and the C'_i given by 12.32 satisfy the conclusions of lemma 12.4.

This completes the proof of lemma 12.4.

Proof of lemma 12.6. All the subcomplexes $\{B'_j : j \geq 0\}$ are equipped with compatible piecewise smooth (triangulation) structures. These can be converted to compatible piecewise linear structures for the $\{C_j : j \geq 0\}$. (The B'_j were defined prior to 12.5.)

Note that it follows from 11.0-11.13 that the $\{C_j : j \geq 0\}$ defined in 12.5 satisfy property 12.2(a). We just saw in the preceding paragraph that the $\{C_j : j \geq 0\}$ of 12.5 satisfy property 12.2(d).

Finally note that it follows from 11.0(b) and from 6.8(d) (as well as from 11.1-11.13) that the $\{C_j : j \geq 0\}$ satisfy 12.2(b)(c).

This completes the proof of lemma 12.6.

13. CONSTRUCTION OF MARKOV CELLS

In this section we prove theorem 1.5 when Hypothesis 5.2 is satisfied. In section 14 we will prove theorem 1.5 when Hypothesis 5.2 is not satisfied.

PROOF OF THEOREM 1.5: Because Hypothesis 5.2 is assumed the Markov cell structure X which is constructed in this proof will satisfy the following property as well as the conclusions of theorem 1.5.

13.0. For each cell $d \in X$ there is a cell $e \in C$ and topological cells d_u, d_s in $X_{u,e}, X_{s,e}$ such that $d = g_e(d_u \times d_s)$. We will call d_u the *unstable factor* of d and call d_s the *stable factor* of d. (Here C comes from 6.2.)

The unstable factor of the cell structure X (that is, the factors d_u of the cells $d \in X$) is constructed from the limit balls $\{{}_kY_j(f;\infty) : k \in I_e,\ e \in C,\ f \in K_{u,e},\ 1 \leq j \leq y\}$ of 12.0.

The stable factor of the cell structure X (that is, the factors d_s of the cells $d \in X$) is constructed from an entirely different set of limit balls which we discuss now.

Define the diffeomorphism $\hat{F} : M \to M$ to be $F^{-1} : M \to M$. Set $\hat{X}_{u,e} = X_{s,e},\ \hat{X}_{s,e} = X_{u,e}$ for all $e \in C$, where $C, X_{u,e}, X_{s,e}$ come from 6.7. If we apply proposition 6.8 to $\hat{F}$, $\{\hat{X}_{u,e} : e \in C\}$ (instead of to F, $\{X_{u,e} : e \in C\}$) we obtain smooth triangulations $\{\hat{K}_{u,e} : e \in C\}$ which satisfy 6.8(a)-(e). If we apply 7.2 to the $\{\hat{K}_{u,e} : e \in C\}$ we obtain redundant almost P.S. ball structures $\{\hat{Y}_i(f) : f \in \hat{K}_{u,e},\ 1 \leq i \leq x\}$ for each $\hat{K}_{u,e}$ which satisfy 7.2(a)-(c). If in the paragraph preceding 8.1 we replace R^m by R^n the construction of that paragraph will yield a collection of balls $\{\hat{B}_{s,i} : i \in \hat{I}_e,\ e \in C\}$ in R^n which satisfy 8.1 (in 8.1(b) we must replace m by n). If we apply lemma 8.2, proposition 8.5, theorem 9.6, and theorem 12.1 to $\hat{F}$, $\{\hat{Y}_i(f) : f \in \hat{K}_{u,e},\ e \in C,\ 1 \leq i \leq x\}$ (instead of applying these results to F, $\{Y_i(f) : f \in K_{u,e},\ e \in C,\ i \leq i \leq x\}$) then we will get collections of balls

$$\{{}_k\hat{Y}_i'(f) : f \in \hat{K}_{u,e},\ k \in \hat{I}_e,\ e \in C,\ 1 \leq i \leq y\}$$
$$\{{}_k\hat{Y}_i(f;\infty) : f \in \hat{K}_{u,e},\ k \in \hat{I}_e,\ e \in C,\ 1 \leq i \leq y\}$$

which satisfy the conclusions of 8.5 and 12.1 (when in 8.5, 12.1 we replace the $\{B_{s,k} : k \in I_e,\ e \in C\}$ by the $\{\hat{B}_{s,k} : k \in \hat{I}_e,\ e \in C\}$).

The stable factor of the cell structure X is constructed from the limit balls $\{{}_k\hat{Y}_j(f;\infty) : k \in \hat{I}_e,\ e \in C,\ f \in \hat{K}_{u,e},\ 1 \le j \le y.\}$

Note that an integer $q > 0$ appears implicitly in each of theorems 9.6 and 12.1 (see 9.1 and the definition of the ${}_kY_i(f;t)$, $1 \le t \le \infty$, given in 9.5, 12.0). We choose $N > 0$ of theorem 1.5 so as to satisfy the following property.

13.1. If the integer q satisfies $q \ge N$, then theorem 12.1 holds for both the $\{{}_kY_i(f) : f \in K_{u,e}, k \in I_e, e \in C,\ 1 \le i \le y\}$ and for the $\{{}_k\hat{Y}_i(f) : f \in \hat{K}_{u,e},\ k \in \hat{I}_e,\ e \in C,\ 1 \le i \le y\}$.

Once an integer q is selected so that $q \ge N$ then we fix the cell structure C of 6.2. We define the neighborhood U for Λ in M of theorem 1.5 as follows.

13.2. $U = |C|$.

We denote by J the collection of all sets $g_e({}_kY_i(f;\infty) \times {}_\ell\hat{Y}_j(g;\infty))$ which satisfy the following property.

13.3. $e \in C$; $k \in I_e$, $\ell \in \hat{I}_e$; $1 \le i,j \le y$; $f \in K_{u,e}$, $g \in \hat{K}_{u,e}$; and ${}_kY_i(f;\infty) \times {}_\ell\hat{Y}_j(g;\infty) \subset \hat{B}_{s,\ell} \times B_{s,k}$.

Note it follows from 6.2(b), 6.7(b), 7.2(b), 8.5(b), 9.5, 11.0(b) that there is a number $\beta > 0$ satisfying the following properties.

13.4. (a) β is independent of q and of the upper bound for the diameters of all the sets in J.

(b) The collection of sets J is a union $\cup_{j=1}^{\beta} J_j$ of pairwise disjoint subsets $J_j \subset J$.

(c) For any $j \in \{1, 2, \dots, \beta\}$ and any sets $A, B \in J_j$ we must have that $A \cap B = \emptyset$.

For each $e \in C$ there is a subset $N'_{s,e} \subset X_{s,e}$ defined by $N'_{s,e} = \cup_{i \in I_e} B_{s,i}$. We let $\hat{N}'_{s,e}$ denote the corresponding subset of $\hat{X}_{s,e}$. In all that follows there will be no loss of generality in assuming that for each $e \in C$ and each $i \in \{1, 2, \dots, \beta + 1\}$ that there is a compact subset ${}_iN'_{s,e} \subset N'_{s,e}$ and a compact subset ${}_i\hat{N}'_{s,e} \subset \hat{N}'_{s,e}$ such that the following properties are satisfied.

13.5. (a) For each $e \in C$ we must have that ${}_1N'_{s,e}, {}_1\hat{N}'_{s,e}$ are arbitrarily large compact subsets of $X_{s,e}, \hat{X}_{s,e}$.

(b) For each $e \in C$ and each $i \in \{1, 2, \dots, \beta\}$ we have that ${}_iN'_{s,e} \subset \mathrm{Int}({}_{i+1}N'_{s,e})$ and that ${}_i\hat{N}'_{s,e} \subset \mathrm{Int}({}_{i+1}\hat{N}'_{s,e})$.

(c) ${}_{\beta+1}N'_{s,e} \subset \mathrm{Int}({}_0\hat{N}_{u,e} \cap {}_0N_{s,e})$ and ${}_{\beta+1}\hat{N}'_{s,e} \subset \mathrm{Int}({}_0N_{u,e} \cap {}_0\hat{N}_{s,e})$, where ${}_0N_{u,e}, {}_0N_{s,e}$ come from 9.0 and ${}_0\hat{N}_{u,e}, {}_0\hat{N}_{s,e}$ are similarly associated to $\hat{X}_{u,e}, \hat{X}_{s,e}$.

For each $i \in \{1, 2, 3, \ldots, \beta + 1\}$ there is a subset $J^i \subset J$ defined as follows. Any $A \in J$ has the form $g_e({}_kY_i(f;\infty) \times {}_\ell\hat{Y}_j(g;\infty))$ as in 13.3. We let $A \in J^i$ if and only if ${}_kY_i(f;\infty) \times {}_\ell\hat{Y}_j(g;\infty)$ intersects ${}_i\hat{N}'_{s,e} \times {}_iN'_{s,e}$. For each $A \in J^1$ and each positive integer k we define a subset $J(A,k) \subset J^{k+1}$ as follows.

<u>13.6.</u> $B \in J(A,k)$ if and only if the following two requirements are meant. Firstly there must exist a sequence $A_1, A_2, \ldots, A_{k+1}$ in J such that $A_1 = A$, $A_{k+1} = B$ and $A_i \cap A_{i+1} \neq \emptyset$ holds for all $i \in \{1, 2, \ldots, k\}$. Secondly we must have that $B \in J^{k+1}$.

For each $A \in J^1$ and any $x \in \{1, 2, \ldots, \beta\}$ we define a partitioning of A—denoted by $A(x)$—as follows. A is a set of the form $g_e({}_kY_i(f;\infty) \times {}_\ell\hat{Y}_j(g;\infty))$ where $e \in C$, $f \in K_{u,e}$, $g \in \hat{K}_{u,e}$. First we define a partitioning of the ball ${}_kY_i(f;\infty)$ denoted by ${}_kY_i(f;\infty)(x)$.

<u>13.7.</u> Let $B_1, B_2, B_3, \ldots, B_t$ denote the images of all the sets in $J(A,x)$ under the composite map

$$\mathrm{Image}(g_e) \xrightarrow{g_e^{-1}} R^n \times R^m \xrightarrow{\text{proj.}} R^n.$$

Then ${}_kY_i(f;\infty)(x)$ is the partitioning of ${}_kY_i(f;\infty)$ generated under intersection by all the sets $\{\overline{B_1 - B_j},\ B_1 \cap B_j,\ \partial B_1,\ B_1 \cap \partial B_j : 1 \le j \le t\}$. Here we assume that $B_1 = {}_k Y_i(f;\infty)$.

Using this same recipe (as in 13.7) we can also define a partitioning of ${}_\ell\hat{Y}_j(g;\infty)$—denoted by ${}_\ell\hat{Y}_j(g;\infty)(x)$—as follows.

<u>13.8.</u> Let $\hat{B}_1, \hat{B}_2, \hat{B}_3, \ldots, \hat{B}_t$ denote the images of all the sets in $J(A,x)$ under the composite map

$$\mathrm{Image}(g_e) \xrightarrow{g_e^{-1}} R^n \times R^m \xrightarrow{\text{proj.}} R^m.$$

Then ${}_\ell\hat{Y}_j(g;\infty)(x)$ is the partitioning of ${}_\ell\hat{Y}_j(g;\infty)$ generated under intersection by all the sets $\{\overline{\hat{B}_1 - \hat{B}_j},\ \hat{B}_1 \cap \hat{B}_j,\ \partial\hat{B}_1,\ \hat{B}_1 \cap \partial\hat{B}_j : 1 \le j \le t\}$. Here we assume that $\hat{B}_1 = {}_\ell\hat{Y}_j(g;\infty)$.

Finally we can define the partitioning $A(x)$ of A as follows.

13.9. $A(x)$ is the image under $g_e : R^n \times R^m \to M$ of the product

$$_kY_i(f;\infty)(x) \times {}_\ell\hat{Y}_j(g;\infty)(x).$$

We claim that the $\{A(x) : A \in J,\ x \in \{1, 2, \ldots, \beta\}\}$ satisfy the following properties.

13.10. (a) Each $A(x)$ is a regular cell structure for A.

(b) For any $A, B \in J^1$ we must have that $A \cap B$ is the underlying set of subcomplexes of $A(1)$ and of $B(1)$.

(c) For any $A, B \in J^1$ and any $x \in \{1, 2, \ldots, \beta\}$ let C_1 denote the cell structure inherited by $A \cap B$ from $A(x)$ and let C_2 denote the cell structure inherited by $A \cap B$ from $B(x+1)$ (see part (b)). Then C_2 is a subdivision of C_1. In particular (if $A = B$) then we get that $A(x+1)$ is a subdivision of $A(x)$.

Properties 13.10(b)(c) are an immediate consequence of the definition of the $A(x)$, provided it is assumed that the diameters of all the sets in J are arbitrarily small.

Towards verifying property 13.10(a) we first note that each B_j in 13.7 is a set of the form $_{k'}Y_{i',e}(f';\infty)$. Thus we may apply lemma 10.21 to conclude that each $_kY_i(f;\infty)(x)$, is a regular cell structure for $_kY_i(f;\infty)$. (Note in order to verify that hypothesis 10.20 holds for this application of lemma 10.21 we must use 12.1, 7.2, 13.4–13.9, and assume that the diameters of the sets in J are arbitrarily small.) A similar application of lemma 10.21 shows that each $_\ell\hat{Y}_j(g;\infty)(x)$ is a regular cell structure for $_\ell\hat{Y}_j(g;\infty)$. Now it follows from 13.9 that $A(x)$ is a regular cell structure for A.

We can now define the cell structure X of theorem 1.5. Note that it follows from 13.10 and 13.4 that a regular cell structure X is uniquely determined by the following requirements.

13.11. (a) $|X| = \cup_{A \in J'} A$, where $J' \subset J^1$ and $A \in J'$ if and only if $A \cap |C| \neq \emptyset$.

(b) Each $A \in J'$ is the underlying set of a subcomplex of X denoted by X_A.

(c) If $A \in J_j \cap J'$ (see 13.4(b) for J_j) then $A(\beta - j + 1)$ is subdivided by X_A.

(d) X is the cell structure with least number of cells which satisfies (a)–(c) above.

It follows from 13.9 that the cell structure X defined by 13.11 must satisfy 13.0. That X satisfies the conclusions of theorem 1.5 follows immediately from 13.11(a), 13.5, 13.2, 13.11, 13.9 and the following claim.

CLAIM 13.12. Let A, A' be any sets in J' and let d be any cell in $A'(\beta)$. Note that A has the form $g_e({}_kY_i(f;\infty) \times {}_\ell\hat{Y}_j(g;\infty))$ for some $e \in C$, $f \in K_{u,e}$, $g \in \hat{K}_{u,e}$, $k \in I_e$, $\ell \in \hat{I}_e$, and $1 \leq i,j \leq y$.

(a) The image of $F^q(d) \cap A$ under the composite map

$$A \xrightarrow{g_e^{-1}} R^n \times R^m \xrightarrow{\text{proj.}} R^n$$

is a subcomplex of ${}_kY_i(f;\infty)(1)$.

(b) The image of $F^{-q}(d) \cap A$ under the composite map

$$A \xrightarrow{g_e^{-1}} R^n \times R^m \xrightarrow{\text{proj.}} R^m$$

is a subcomplex of ${}_\ell\hat{Y}_j(g;\infty)(1)$.

Thus to complete the proof of theorem 1.5 it remains only to verify claim 13.12. We shall verify 13.12(a). The same argument will work to also verify 13.12(b). Note that A' has the form $g_{e'}({}_{k'}Y_{i'}(f';\infty) \times {}_{\ell'}\hat{Y}_{j'}(g';\infty))$ for some $e' \in C$, $f' \in K_{u,e'}$, $g' \in \hat{K}_{u,e'}$, $k' \in I_{e'}$, $\ell' \in \hat{I}_{e'}$, $1 \leq i',j' \leq y$. Note also that $d = g_{e'}(d_u \times d_s)$ where $d_u \in {}_{k'}Y_{i'}(f';\infty)(\beta)$ and $d_s \in {}_{\ell'}\hat{Y}_{j'}(g';\infty)(\beta)$. By definition of ${}_{k'}Y_{i'}(f';\infty)(\beta)$ there must exist a finite set of balls $\{{}_{k_z}Y_{i_z,e'}(f_z;\infty); z \in I\}$ satisfying the following properties.

<u>13.13.</u> (a) $d_u = U \cap (\cap_{z \in I} U_z)$, where U is any one of the sets ${}_{k'}Y_{i'}(f';\infty)$ or $\partial_{k'}Y_{i'}(f';\infty)$, and <u>$U_z$ denotes any one of the sets</u> $\{{}_{k_z}Y_{i_z,e'}(f_z;\infty), \partial_{k_z}Y_{i_z,e'}(f_z;\infty)$ or $\overline{{}_{k'}Y_{i'}(f';\infty) - {}_{k_z}Y_{i_z,e'}(f_z;\infty)}$.

(b) For each $z \in I$ there is a set $B_z \in J(B,\beta)$ such that ${}_{k_z}Y_{i_z,e'}(f_z;\infty)$ is equal to the image of B_z under the composite map

$$\text{Image}(g_{e_z}) \xrightarrow{g_{e_z}^{-1}} R^n \times R^m \xrightarrow{\text{proj.}} R^n.$$

(Here $e_z \in C$, $f_z \in K_{u,e_z}$, $k_z \in I_{e_z}$, and we also may assume that $B_z \subset \text{Image}(g_{e_z})$.)

Note that it follows from 9.0, 9.1, 9.5, 12.0, 13.3, 13.5, 11.0(b)(c) that if the sets in J are chosen to have arbitrarily small diameters then the following property must hold.

<u>13.14.</u> The image of each $(F^q \circ g_{e_z}(B_z)) \cap A$ under the composite map

$$A \xrightarrow{g_e^{-1}} R^n \times R^m \xrightarrow{\text{proj.}} R^n$$

is the underlying set of a subcomplex of ${}_kY_i(f;\infty)(1)$. Now note that 13.12(a) is implied by 13.11, 13.13, and 13.14.

This completes the proof of theorem 1.5.

14. REMOVING THE FOLIATION HYPOTHESIS

In this section we discuss the modifications that must be made to the proof of theorem 1.5 when hypothesis 5.2 does not hold. The simple idea behind all such modifications is that 5.1(a) is esentially a tangential version of hypothesis 5.2. Therefore we try to make do with 5.1(a) in place of 5.2.

The first modification occurs in the selection of charts $g_i : R^n \times R^m \to M$, $i \in I$, of 6.1. Since 5.2 does not necessarily hold we can not seslect the g_i to satisfy 6.1(a). Instead we explicitly describe some charts (in terms of the geometry of M) which satisfy an approximate tangential version of 6.1(a).

Let V, $\overline{\xi}_u$, $\overline{\xi}_s$ be as in 5.1 and 5.3. For each $p \in V$ choose $h_p : R^n \times R^m \to T(V)_p$ to be a linear isometry such that $h_p(R^n \times 0) = \overline{\xi}_u|_p$ and $h_p(0 \times R^m) = \overline{\xi}_s|_p$. Choose $\delta > 0$ sufficiently small so that the map $g_p : B^n_\delta \times B^m_\delta \to M$ defined by 14.1(a) satisfies 14.1(b), where B^n_δ, B^m_δ denote the balls of radius δ centered at the origin in R^n, R^m.

<u>14.1.</u> (a) Define g_p to be the composite map

$$B^n_\delta \times B^m_\delta \xrightarrow{h_p} T(V)_p \xrightarrow{\exp} M,$$

where exp:$T(M) \to M$ is the exponential map.

(b) $g_p : B^n_\delta \times B^m_\delta \to M$ is a well defined one-one smooth immersion.

Note that there is a continuous map $r : [0,1] \to [0,\infty)$ with $r(0) = 0$ such that g_p of 14.1 also satisfy the following properties.

<u>14.2.</u> (a) The map $r : [0,1] \to [0,+\infty)$ depends only on the geometry of M.

(b) For any $v \in T(B^n_\delta \times B^m_\delta)$ we have that

$$(1 - r(\delta))\langle v, v\rangle \leq \langle dg_p(v), dg_p(v)\rangle_M \leq (1 + r(\delta))\langle v, v\rangle,$$

where $\langle,\rangle$ denotes the usual Riemannian tensor on $R^n \times R^m$ and $\langle,\rangle_M$ denotes the Riemannian tensor on M.

(c) For any $p, q \in V$ let $X_{p,q}$ denote the interior of the domain of the composite map $g_q^{-1} \circ g_p$. There are smooth maps $I_1 : R^n \to R^n$, $I_2 : R^m \to R^m$, $v : R^n \times R^m \to R^n$, and $u : R^n \times R^m \to R^m$ which satisfy the following properties.

(i) For any $(x, y) \in X_{p,q}$—with $x \in B^n_\delta$ and $y \in B^m_\delta$—we must have that $g_q^{-1} \circ g_p(x, y) = (I_1(x) + v(x, y), I_2(y) + u(x, y))$.

(ii) I_1 and I_2 are isometries.
(iii) Both $v : R^n \times R^m \to R^n$ and $u : R^n \times R^m \to R^m$ are a distance less than $r(\delta)\delta$ from the zero map in the C^0-metric, and a distance less than $r(\delta)$ from the zero map in the C^1-metric.

We leave the proof of the following lemma to the reader.

LEMMA 14.3. *For δ sufficiently small in 14.1 there will exist a finite subset $I \subset V$ such that the maps $g_i : B^n_\delta \times B^m_\delta \to M$, $i \in I$, of 14.1 and 14.2 satisfy the following additional properties.*

(a) There is a number $\alpha > 0$, which depends only on $\dim(M)$, such that any one of the sets $\{\text{Image}(g_i) : i \in I\}$ intersects at most α other such sets. Moreover $\text{Image}(g_i) \subset V$ holds for all $i \in I$.

(b) Set $U = \cup_{i \in I}\, g_i(B^n_{a_i} \times B^m_{b_i})$, where $a_i, b_i \in (\delta' - 10^{-2}\delta', \delta')$ and $\delta' = 10^{-2}\delta$. Then U is a sufficiently good approximation to V so that

$$F^q(F^q(U) \cap \partial U) \subset M - V,$$
$$F^{-q}(F^{-q}(U) \cap \partial U) \subset M - V$$

both hold, where ∂U denotes the topological boundary for U in M. (Compare with 5.3.) Moreover U is a neighborhood for Λ.

The $\{g_i : i \in I\}$ of 14.3 will take the place of the $\{g_i : i \in I\}$ of 6.1.

Now we proceed with the modifications in the construction of the cell structure C of 6.2. The fact of the matter is that because of the failure of 5.2 to hold we now require a lot better metric control in C than is stated in 6.2. The desired metric control for C is most efficiently stated in terms of the $M_{u,i}, K_{u,i}, M_{s,i}, K_{s,i}$ of 6.3, 6.4.

LEMMA 14.4. *Let N_1 denote a very large given positive number and let $r_1 : [0,1] \to [0,\infty)$ be the map defined by $r_1(t) = N_1 r(t)$, where $r : [0,1] \to [0,\infty)$ comes from 14.2. Then for sufficiently small δ in 14.1 we can choose smooth submanifolds $M_{u,i}, M_{s,i}$ in B^n_δ, B^m_δ, and we can choose smooth triangulations $K_{u,i}, K_{s,i}$ of the $M_{u,i}, M_{s,i}$ so that the following properties hold.*

(a) $M_{u,i} = B^n_{a_i}$ and $M_{s,i} = B^m_{b_i}$ for some numbers a_i, b_i in $(\delta' - 10^{-2}\delta', \delta')$ with $\delta' = 10^{-2}\delta$.

(b) $K_{u,i}$ is the intersection with $M_{u,i}$ of a linear triangulation $T_{u,i}$ in B^n_δ such that $|T_{u,i}| \supset M_{u,i}$ and $T_{u,i}$ intersects $\partial M_{u,i}$ transversely. Likewise $K_{s,i}$ is the intersection with $M_{s,i}$ of a linear triangulation $T_{s,i}$ in B^m_δ such that $|T_{s,i}| \supset M_{s,i}$ and $T_{s,i}$ intersects $\partial M_{s,i}$ transversely.

(c) For a fixed $i \in I$ define a partition $L_{u,i}$ of a region of B^n as follows. For each $j \in I$ such that $g_j(B^n_{\delta'} \times B^m_{\delta'}) \cap g_i(B^n_{\delta'} \times B^m_{\delta'}) \neq \emptyset$ choose a point

$q_j \in M_{s,j}$ and let K_{u,i,j,q_j} denote the image of the smooth triangulation $K_{u,j} \times (q_j)$ under the composite map

$$B^n_{\delta'} \times B^m_{\delta'} \xrightarrow{g_j} M \xrightarrow{g_i^{-1}} B^n_\delta \times B^m_\delta \xrightarrow{\text{proj.}} B^n_\delta.$$

We require that the smooth triangulations $\{K_{u,i,j,q_j} : j \in I\}$ must be in transverse position to one another and that $L_{u,i}$ is the partition of the region $\cup_{j \in I} |K_{u,i,j,q_j}|$ generated by the $\{K_{u,i,j,q_j} : j \in I\}$ Moreover we require that every partition set in $L_{u,i}$ be a P.S. cell.

(d) For each $i \in I$ we can define a partition $L_{s,i}$ of a region of B^m_δ by using the recipe in part (c) but with the roles of the B^m_δ and B^n_δ reversed. Again we require that the $\{K_{s,i,j,q_j} : j \in I\}$ must be in transverse position to one another—where $q_j \in M_{u,j}$—and that each partition set in $L_{s,i}$ is a P.S. cell.

(e) There is a number $\gamma_1 \in (0,1)$ which depends only on $\dim(M)$. The following constraints are satisfied by the thicknesses and the diameters of the $L_{u,i}, L_{s,i}, K_{u,i,j,q_j}, K_{s,i,j,q_j}$ of (c) and (d). Let X denote any of the partitions mentioned in the preceding sentence. Then we must have:

$$\gamma_1 < \tau(X),$$
$$\gamma_1 r_1(\delta)\delta < \overline{D}(X) < r_1(\delta)\delta.$$

PROOF OF LEMMA 14.4: We begin with linear triangulations $T_{u,i}$ and $T_{s,i}$ of very large regions in B^n_δ and B^m_δ, which satisfy:

14.5

$$\gamma r_1(\delta)\delta < \overline{D}(T_{u,i}) < \frac{1}{2} r_1(\delta)\delta,$$
$$\gamma r_1(\delta)\delta < \overline{D}(T_{s,i}) < \frac{1}{2} r_1(\delta)\delta,$$
$$\tau(T_{u,i}) > \gamma,$$
$$\tau(T_{s,i}) > \gamma,$$

for some $\gamma \in [0,1]$ which depends only on $\dim(M)$.

Note it follows from 14.3(a) that there are subsets $I_j \subset I$, for $j \in \{1, 2, \ldots, \alpha^2 + \alpha\}$, which satisfy the following property.

14.6. (a) $I = \bigcup_{j=1}^{\alpha^2+\alpha} I_j;\ I_j \cap I_{j'} = \emptyset$ if $j \neq j'$.

(b) For any $j \in \{1, 2, \ldots, \alpha^2 + \alpha\}$ if $i_1, i_2 \in I_j$ and $i_1 \neq i_2$ then there is no $k \in I$ such that Image(g_k) intersects both Image(g_{i_1}) and Image(g_{i_2}).

Now the proof proceeds by induction over the sequence $I_1, I_2, I_3, \ldots$. For our induction hypothesis we assume the conclusions of 14.4 hold true when the $L_{u,i}$ and $L_{s,i}$ of 14.4(c)(d) are constructed by using only those K_{u,i,j,q_j} and K_{s,i,j,q_j} with $j \in \cup_{k=1}^{r} I_k$. We let the number ${}_r\gamma$ denote the number γ_1 of 14.4(e) which works at this stage of the induction argument. We claim that by moving each $T_{u,i}$ and $T_{s,i}$ (with $i \in \bigcup_{k=1}^{r+1} I_k$) through a P.L. flow (see 2.3 for the definition of a P.L. flow) and changing a_i, b_i a little bit we can conclude the induction step in this argument for some ${}_{r+1}\gamma$ which depends only on $\dim(M)$. The part of this claim involving the P.L. flow of the $T_{u,i}$ and the $T_{s,i}$ follows from lemma 3.1 (and from the P.S. version of lemma 3.1). The remaining details in the induction step are left to the reader.

Note that the γ_1 that finally works in 14.4(e) is equal to the ${}_{\alpha^2+\alpha}\gamma$ obtained through our induction argument. Thus it is essential that α depend only on $\dim(M)$ (see 14.3(a)) in order that γ_1 may depend only on $\dim(M)$ as claimed in 14.4(e).

This completes the proof of lemma 14.4.

Our plan now is to use 14.4 and 14.2 (which the $\{g_i : i \in I\}$ satisfy) to define the cell complexes $S(e_u \times e_s)$ of 6.6, and then to use 6.6 to define the cell complex C. Note that the old definition of the $S(e_u \times e_s)$ depended on the existence of rectangular charts as in 6.1(a), which is just a local version of Hypothesis 5.2. Since 5.2 is not now in effect we must devise a new definition of the $S(e_u \times e_s)$.

For any $i \in I$ and any $e_u \in K_{u,i}$, $e_s \in K_{s,i}$ we define a number of smooth polyhedra structures for $g_i(e_u \times e_s)$. Let $J \subset I$ denote the subset of all $j \in I$ such that there is $e'_u \in K_{u,i}$, $e'_s \in K_{s,i}$ (which depend on j and satisfy $e_u \subset e'_u$, $e_s \subset e'_s$) so that $g_i(e'_u \times e'_s) \cap g_j(B^n_{a_i} \times B^m_{b_i}) \neq \emptyset$. For each $j \in J$ let P_j denote the partitioning of $B^n_\delta \times B^m_\delta$ into the following sets: $e_1 \times e_2$, where $e_1 \in K_{u,j}$, $e_2 \in K_{s,j}$; $\overline{(B^n_\delta - B^n_{a_i})} \times e_2$, where $e_2 \in K_{s,j}$; $e_1 \times \overline{(B^m_\delta - B^m_{b_i})}$, where $e_1 \in K_{u,j}$; $\overline{(B^n_\delta - B^n_{a_i})} \times \overline{(B^m_\delta - B^m_{b_i})}$. For each $j \in J$ let Q_j denote the partitioning of $g_i(e_u \times e_s)$ into sets of the form $g_i(e_u \times e_s) \cap g_j(Y)$, where Y is one of the partition sets in P_j. Finally we define $S(e_u \times e_s)$ to be the partitioning of $g_i(e_u \times e_s)$ generated by all the $\{Q_j : j \in J\}$ (see 2.5). We note that it follows from 14.2 and 14.4 that $S(e_u \times e_s)$ is a smooth polyhedral structure for $g_i(e_u \times e_s)$ each partition set of which is a P.S. cell, provided δ of 14.1 is chosen small enough.

Note we can use 6.6 to define a cell structure C. Note that we need to use 14.2 and 14.4 once again in order to verify that there is a cell structure C well defined by 6.6.

The following list of properties of C should replace the list found in 6.2.

14.7. (a) For each $e \in C$ we must have that

$$1/2\gamma_1 r_1(\delta)\delta < \text{diameter}(e) < 2r_1(\delta)\delta.$$

(b) $|C| = U$. (Compare with 6.2(b).)

(c) $N(C) < b$, where b depends only on $\dim(M)$. (Compare with 6.2(c).)

We come now to the selection of the charts $g_e : X_{u,e} \times X_{s,e} \to M$, $e \in C$, of 6.7. Again, because hypothesis 5.2 is not assumed, the charts $\{g_e : e \in C\}$ must satisfy some strong metric conditions as well as 6.7(a)(b). All of these conditions are contained in the next lemma.

LEMMA 14.8. *The charts $g_e : X_{u,e} \times X_{s,e} \to M$, $e \in C$, can be chosen to satisfy the following properties.*

(a) For each $e \in C$ we have that $X_{u,e} \subset B^n_{\delta'}$, $X_{s,e} \subset B^m_{\delta'}$, and $g_e = g_i | X_{u,e} \times X_{s,e}$ for some $i \in I$, where I and $g_i : B^n_\delta \times B^m_\delta \to M$ are as in 14.2–14.4.

(b) 6.7(a)(b) are satisfied.

(c) Let N_2 be a given very large positive number, and let $r_2 : [0,1] \to [0,\infty)$ be the map defined by $r_2(t) = N_2 r(t)$, where $r : [0,1] \to [0,\infty)$ comes from 14.2. Then for sufficiently small δ in 14.1 the following will hold true. Let $X'_{u,e}$ denote all points in $X_{u,e}$ which are a distance greater than $r_2(\delta)\delta$ from $\overline{B^n_\delta - X_{u,e}}$, where δ comes from 14.1 and 14.2. A subset $X'_{s,e} \subset X_{s,e}$ is defined in the same way. Then we require that for any $e \in C$ if we replace $X_{u,e}$, $X_{s,e}$ in 6.7 by the $X'_{u,e}$, $X'_{s,e}$ then 6.7(a)(b) still remain true.

PROOF OF LEMMA 14.8. The proof is an application of 14.2, 14.4, and of the definition of C. The relation of our present large number N_2, and the large number N_1 of 14.4 is as follows: first N_2 is given, and then N_1 must be selected to be much larger than N_2. The remaining details are left to the reader.

This completes the proof of lemma 14.8.

REMARK 14.9. Note that the charts $g_i : B^n_\delta \times B^m_\delta \to M$ of 14.4 satisfy 14.1 and 14.2 (in fact I is a subset of V). Thus, by 14.8, the charts $g_e : X_{u,e}, \times X_{s,e} \to M$ must also satisfy the conditions 14.2(a)–(c).

Having decided upon which properties the charts $g_e : X_{u,e}, \times X_{s,e} \rightarrow M$, $e \in C$, should satisfy when hypothesis 5.2 does not hold, we must now address the problem of how to formulate propositions 6.8, 8.5 and lemma 7.2 when hypothesis 5.2 does not hold.

First we describe how to reformulate proposition 6.8. The first problem apparent in the present formulation of 6.8 is that the smooth triangulations $K_{u,e,e'}$, for $e, e' \in C$, are not well defined when hypothesis 5.2 is not satisfied. To overcome this difficulty we introduce smooth triangulations $K_{u,e,e',q}$ for each $e, e' \in C$ and each $q \in X_{s,e'}$. Recall that $g_e : X_{u,e} \times X_{s,e} \rightarrow M$ is equal to the restriction $g_i|X_{u,e} \times X_{s,e}$, where $g_i : B_\delta^n \times B_\delta^m \rightarrow M$ is one of the charts of 14.3. Define $K_{u,e,e',q}$ to be equal to the image of the triangulation $K_{u,e'} \times (q)$ under the composite map

$$X_{u,e'} \times X_{s,e'} \xrightarrow{g_{e'}} M \xrightarrow{g_i^{-1}} B_\delta^n \times B_\delta^m \xrightarrow{\text{proj.}} B_\delta^n,$$

provided that either $e \subset e'$ or that $e' \subset e$ must hold. If neither $e \subset e'$ nor $e' \subset e$ hold then we set $K_{u,e,e',q}$ equal the empty triangulation. Another problem which occurs in the present formulation of 6.8 is that 6.8(b) cannot possibly be true when in 6.8(b) we let $K_{e'}$ stand for $K_{u,e,e',q}$ for some fixed but arbitrary point q in $X_{s,e'}$. (The reason is that in general as $q \in X_{s,e'}$ varies we get a smooth family $\{K_{u,e,e',q} : q \in X_{s,e'}\}$ of smooth triangulations; generally speaking this will be an m-dimensional family.) To overcome this difficulty we need the following definition.

Definition 14.10. *For any $e \in C$ let B be as in 6.8(b). For each $e' \in C$ choose a point $q \in X_{s,e'}$ and set $K_{e'} = K_{u,e,e',q}$. Let $\varepsilon, \varepsilon' > 0$ be given numbers. Let K_B denote the collection $\{K_{e'} : e' \in B\}$. We say that the collection of smooth full polyhedral K_B are in B-***transverse position*** to one another* **modulo** *$(\varepsilon, \varepsilon')$ if the following properties hold.*

(a) For each $K_{e'}$ in K_B there is a piecewise smooth flow of $K_{e'}$ (see 4.10) consisting of a linear parametrization $(h_{e'}, L_{e'})$ for $K_{e'}$ and an isotopy $\varphi_{e',t} : R^n \rightarrow R^n$, $t \in [0,1]$, of $h_{e'} : R^n \rightarrow R^n$, all of which satisfy (b).

(b) For each triangle $d \in L_{e'}$ and each $t \in [0,1]$ we must have that the restriction $\varphi_{e',t}|d$ is a distance less than ε from the inclusion map $d \subset R^n$ with respect to the C^0 metric, and a distance less than ε' from the inclusion map $d \subset R^n$ with respect to the C^1-metric.

(c) Let K'_B denote the collection $\{K'_{e'} : e' \in B\}$, where $K'_{e'}$ is the image of $L_{e'}$ under the map $\varphi_{e',1} : R^n \rightarrow R^n$. Then the collection K'_B are in B-transverse position to one another (as described in 4.5).

Here is our reformulation of proposition 6.8.

PROPOSITION 14.11. *Let $r_0 < r_1 < r_2 < \cdots < r_{n+m}$ be a given increasing sequence of positive integers. Let N_3 be a given very large positive number, and define a map $r_3 : [0,1] \to [0,\infty)$ by $r_3(t) = N_3 r(t)$ where $r : [0,1] \to [0,\infty)$ comes from 14.2. Then for sufficiently small δ in 14.1 there will be linear triangulations $\{K_{u,e} : e \in C\}$ within the $\{X_{u,e} : e \in C\}$ which satisfy the following properties.*

(a) For any $e \in C$, $p \in X_{u,e}$ the distance from p to $|K_{u,e}|$ is less than $\gamma_2 r_3(\delta)\delta$, where $\gamma_2 > 1$ depends only on dim(M) and on r_{n+m}. (Compare with 6.8(a).)

(b) For any $e \in C$ let B be as in 6.8(b). For each $e' \in B$ choose $q \in X_{s,e'}$ and set $K_{e'} = K_{u,e,e',q}$. Let K_B denote the collection $\{K_{e'} : e' \in B\}$. Then the collection of smooth polyhedra K_B must be in B-transverse position to one another modulo $(\gamma_2 r(\delta)\delta, (\gamma_2 r(\delta))$, where $r(\ )$ comes from 14.2. (Compare with 6.8(b).)

(c) By 14.10 and 14.11(b) we can obtain another collection K'_B which are related to the polyhedra of K_B by 14.10(b)(c) where $\varepsilon = \gamma_2 r(\delta)\delta$ and $\varepsilon' = \gamma_2 r(\delta)$ in 14.10. Then for any $i \in \{0,1,2,\ldots,n+m\}$ there is the subcollection $K'_{B^i} \subset K'_B$ associated to the i-dimensional skeleton B^i of B. If in 6.8(c) we replace the K_{B^i} by any K'_{B^i} then 6.8(c) must remain true.

(d) For each $e \in C^i - C^{i-1}$ and for B as in 6.8(b) there is a linear triangulation $L_{u,e}$ in $X_{u,e}$ and a linear r_i-derived subdivision $L_{u,e}^{(r_i)}$ of $L_{u,e}$ such that $K_{u,e} = L_{u,e}^{(r_i)}$. Let $K'(\partial e)$ denote the smooth polyhedron generated by all the $\{K'_{e'} : e' \in \partial e\}$. Let $L'_{u,e}$ denote the image of $L_{u,e}$ under the map $\varphi_{e,1} : R^n \to R^n$ of 14.10. Note that $L'_{u,e}$ is in general not a smooth triangulation in R^n but is a piecewise smooth triangulation in R^n. We require that there are subcomplexes $\hat{L}'_{u,e}, \hat{K}'(\partial e)$ of $L'_{u,e}, K'(\partial e)$ such that

$$|\hat{L}'_{u,e}| = |\hat{K}'(\partial e)| = |L'_{u,e}| \cap |K'(\partial e)|,$$

and such that $\hat{L}'_{u,e}$ subdivides $\hat{K}'(\partial e)$. (Compare with 6.8(d).)

(e) For each $e \in C$ we must have that

$$r_3(\delta)\delta < \overline{D}(L_{u,e}) < \gamma_2 r_3(\delta)\delta,$$
$$r_3(\delta)\delta < \overline{D}(K_{u,e}) < \gamma_2 r_3(\delta)\delta.$$

PROOF OF PROPOSITION 14.11. The proof is by induction and follows the pattern of the proof of 6.8. The only deviation from the proof of 6.8 occurs because 6.8(b) is not required to hold on the nose but rather it is required to hold modulo $(\varepsilon, \varepsilon')$ in the sense of 14.11 and 14.10. Thus the induction step in the proof of 14.11 is not simply an application of propositions 2.10 and 4.8 (as in the induction step in the proof of 6.8) but rather it consists of an application of modulo $(\varepsilon, \varepsilon')$ versions of 2.10 and 4.8 of which we give a rough description now. In the hypothesis of 2.10 and 4.8 we no longer require that the collection K_A be in B-transverse position to one another, but rather require that the collection K_A be in B-transverse position modulo $(\varepsilon, \varepsilon')$ for some small numbers $\varepsilon, \varepsilon' > 0$. The conclusion then of this new version of 2.10 and 4.8 is that the collection K_B (which extends K_A) is in B-transverse position modulo (α, α'), where $\lim_{\varepsilon+\varepsilon'\to 0} \alpha + \alpha' = 0$.

There must be the following relation between the very large numbers N_1, N_2, N_3 of 14.4, 14.8, and 14.11; first we are given N_3; then we must choose N_2 to be much larger than N_3; finally we must choose N_1 to be much larger than N_2.

The remaining details in the proof of 14.11 are left to the reader.

This completes the proof of proposition 14.11.

We would like to generalize many definitions in the sense that B-transversality has been generalized in 14.10. This is the purpose of the next general definition.

DEFINITION 14.12. *Let $K_1, K_2, \ldots, K_k$ be a collection of smooth triangulations in R^n, and let $\varepsilon, \varepsilon' > 0$ be two small numbers. We will say that* **a property** P **holds true modulo** $(\varepsilon, \varepsilon')$ *for the collection $\{K_i : 1 \leq i \leq k\}$ if the following are true.*

(a) For each $i \in \{1, 2, \ldots, k\}$ there is a piecewise smooth flow of K_i consisting of a linear parametrization (h_i, L_i) for K_i and an isotopy $\varphi_{i,t} : R^n \to R^n$, $t \in [0,1]$, of $h_i : R^n \to R^n$.

(b) For each triangle $d \in L_i$ and each $t \in [0,1]$ we must have that the restriction $\varphi_{i,t}|d$ is a distance less than ε from the inclusion map $d \subset R^n$ with respect to the C^0 metric, and is a distance less than ε' from the inclusion map $d \subset R^n$ with respect to the C^1 metric.

(c) Let K_i' denote the image of L_i under the map $\varphi_{i,1} : R^n \to R^n$. Then the collection $\{K_i' : 1 \leq i \leq k\}$ must satisfy property P.

Before stating our reformulation of lemma 7.2 we need some more notation. For each $e \in C$ let $\{Y_i(f) : f \in K_{u,e}, 1 \leq i \leq x\}$ be a redundant ball structure of $K_{u,e}$. Then for each $e, e' \in C$, $f \in K_{u,e'}$, and $q \in X_{s,e'}$ let $Y_{i,e,q}(f'), f'$ denote the images of $Y_i(f) \times (q)$ and $f \times (q)$ under the composite map

$$X_{u,e'} \times X_{s,e'} \xrightarrow{g_{e'}} M \xrightarrow{g_i^{-1}} B_\delta^n \times B_\delta^m \xrightarrow{\text{proj.}} B_\delta^n,$$

provided that either $e \subset e'$ or $e' \subset e$ hold. If neither $e \subset e'$ nor $e' \subset e$ hold then set $Y_{i,e,q}(f') = \emptyset$. Here $g_e = g_i | X_{u,e} \times X_{s,e}$.

Here is our reformulation of lemma 7.2 that is true even when hypothesis 5.2 is not assumed to hold.

LEMMA 14.13. *Given a positive integer x there is for each $e \in C$ a redundant ball structure $\{Y_i(f) : f \in K_{u,e'}, 1 \leq i \leq x\}$ for $K_{u,e}$ such that the following properties hold when δ of 14.1 is sufficiently small.*

(a) *For each $e' \in C$ we randomly choose a point $q(e')$ in $X_{s,e'}$. Then if in 7.2(a)(b) we replace the $Y_{i,e}(f)$ and $K_{u,e,e'}$ by $Y_{i,e,q(e')}(f)$ and $K_{u,e,e',q(e')}$ then all the properties listed in 7.2(a)(b) hold modulo $(\varepsilon, \varepsilon')$, where*

$$\varepsilon = \gamma_3 r(\delta)\delta,$$
$$\varepsilon' = \gamma_3 r(\delta).$$

Here $\gamma_3 > 0$ depends only on $\dim(M)$ and r_{n+m} of 14.11, and $r(\ )$ comes from 14.2.

(b) *If in 7.2(c) we replace the K_{u,e,e_i}, $Y_{j,e}(f)$, $Y_{j',e}(f')$ by $K_{u,e,e_i,q(e_i)}$, $Y_{j,e,q(e_t)}(f)$, $Y_{j',e,q(e_{t'})}(f')$—where $f \in K_{u,e,e_t,q(e_t)}$ and $f' \in K_{u,e,e_{t'},q(e_{t'})}$—then the conclusions of 7.2(c) must remain valid.*

(c) *The conclusions of 7.2(d)(e) hold true if in 7.2(d)(e) we replace the K_{u,e,e_i} and the $Y_{j,e}(f)$ by $K_{u,e,e_i,q(e_i)}$ and by $Y_{j,e,q(e_t)}(f)$, where $f \in K_{u,e,e_t,q(e_t)}$.*

Before reformulating proposition 8.5 we must first slightly modify the properties which we wish the balls $B_{s,i}$ of 8.1 to satisfy when 5.2 is not satisfied.

14.14. (a) The balls $\{B_{s,i} : i \in I_e,\ e \in C\}$ satisfy 8.1(a)(b)(d).

(b) σ of 8.1(a) must satisfy $\sigma = N_4 r(\delta)\delta$, where N_4 is a very large positive number.

(c) N_4 must be very much larger than N_3 given in 14.11.

(d) Let $N'_{s,e} = \cup_{i \in I_e} B_{s,i}$, and let p be any point in $X_{s,e}$. Then the distance from p to $N'_{s,e}$ must be less than 2σ. (Compare with 8.1(c).)

We must also slightly modify the properties that the subsets ${}_jN_{u,e} \subset X_{u,e}$ of 8.4 satisfy.

14.15. (a) $e \subset \operatorname{Int}(\cup_{e' \in e}\, g_{e'}({}_1N_{u,e'} \times N'_{s,e'}))$.

(b) For any $j \in \{1, 2, \ldots, n\}$ and any point p in ${}_jN_{u,e}$ the distance from p to $X_{u,e} - {}_{j+1}N_{u,e}$ is greater than $N_5 r(\delta)\delta$. Moreover if p is in the topological boundary of ${}_jN_{u,e}$ then the distance from p to $X_{u,e} - {}_{j+1}N_{u,e}$ must be less than $2N_5 r(\delta)\delta$. Here N_5 is a very large positive number which is very much larger than the number N_3 of 14.11.

(c) ${}_{n+1}N_{u,e} = |K_{u,e}|$.

Here is our reformulation of proposition 8.5.

PROPOSITION 14.16. *If q, and the r_i of 6.8 and 14.11, are chosen sufficiently large and the δ of 14.1 is chosen sufficiently small then for each $e \in C$, $f \in K_{u,e}$, $i \in \{1,2,\ldots,y\}$, and each $k \in I_e$, there is a subset ${}_kY_i'(f) \subset X_{u,e}$ satisfying all the following properties.*

(a) Property 8.5(a) must hold.

(b) Property 8.5(b) must hold if in 8.5(b) we replace the ${}_{k'}Y_{i,e}(f)$ and ${}_{k'}Y_{i,e}'(f)$ by the images of ${}_{k'}Y_i(f) \times (p)$ and ${}_{k'}Y_i'(f) \times (p)$ under the composite map

$$X_{u,e'} \times X_{s,e'} \xrightarrow{g_{e'}} M \xrightarrow{g_i^{-1}} B_\delta^n \times B_\delta^m \xrightarrow{\text{proj.}} B_\delta^n.$$

Here $g_e = g_i | X_{u,e} \times X_{s,e}$, $k' \in I_{e'}$, $f \in K_{u,e'}$, and p is a fixed but arbitrary point of $X_{s,e'}$. (For future reference we denote the images of the sets ${}_{k'}Y_i(f) \times (p)$ and ${}_{k'}Y_i'(f) \times (p)$ under the above composite map by ${}_{k'}Y_{i,e,p}(f)$ and ${}_{k'}Y_{i,e,p}'(f)$.

(c) Given any $e, e', \in C$, $k \in I_e$, $i \in \{1,2,\ldots,y\}$, and $f \in K_{u,e}$, suppose that $F^q(g_e({}_kY_i'(f) \times B_{s,k}) \subset g_{e'}({}_1N_{u,e'} \times N_{s,e'}')$. Then for any $p \in B_{s,k}$ we have that $\rho_1 \circ g_{e'}^{-1} \circ F^q \circ g_e({}_kY_i'(f) \times (p))$ is a subcomplex of $K_{u,e'}$ modulo $(\varepsilon, \varepsilon')$. Here $\rho_1 : B_\delta^n \times B_\delta^m \to B_\delta^n$ is the projection onto the first factor and $\varepsilon, \varepsilon'$ are as in 14.13(a).

PROOF OF 14.13 AND 14.16. These are just <u>modulo $(\varepsilon, \varepsilon')$</u> versions of the proofs of lemma 7.2 and proposition 8.5. The details are left to the reader.

This completes the proof of lemma 14.13 and proposition 14.16.

We come now to the modifications which must be made in the thickening theorem 9.6 and in the limit theorem 12.1.

The subsets ${}_0N_{u,e} \subset X_{u,e}$, ${}_0N_{s,e} \subset X_{s,e}$ must now satisfy (in place of 9.0(a)(b)) the following property.

<u>14.17.</u> (a) For any point $p \in {}_0N_{u,e}$ the distance from p to $X_{u,e} - {}_1N_{u,e}$ must be greater than $N_5 r(\delta)\delta$. Moreover if p is in the topological boundary of ${}_0N_{u,e}$ then the distance from p to $X_{u,e} - {}_1N_{u,e}$ is less than $2N_5 r(\delta)\delta$. (See 14.15 for ${}_1N_{u,e}$ and N_5.)

(b) For any point $p \in {}_0N_{s,e}$ the distance from p to $X_{s,e} - N_{s,e}'$ must be greater than 10σ. Moreover if p is in the topological boundary of ${}_0N_{u,e}$ then the distance from p to $X_{s,e} - N_{s,e}'$ must be less than 20σ. (See 8.1 and 14.14 for $N_{s,e}'$ and σ.)

Note that 9.1(b) is only true modulo $(\varepsilon, \varepsilon')$ where ε and ε' come from 14.13(a) (compare with 14.16(c)). However there is a *unique* subcomplex $K(k,j,f;k')$ in $K_{u,e'}$ of 9.1(b) for which 9.1(b) is true modulo $(\varepsilon, \varepsilon')$. The uniqueness of the $K(k,j,f;k')$ follows from 14.11(e)(b) and 14.13(a) provided the N_3, γ_2, γ_3 of 14.11 and 14.13 are related as follows.

14.18. The N_3 is chosen to be much larger than either the γ_2 or the γ_3.

In all that follows we assume that 14.18 is in effect.

For each ball ${}_kY_j(f)$ we must modify somewhat the notion of the t'th level thickening ${}_kY_j(f;t)$ given in 9.5. Instead of trying to define the thickening ${}_kY_j(f;t)$—which the failure of 5.2 and 6.1(a) makes difficult—we shall define the t'th level thickening of $g_e({}_kY_j(f) \times {}_5B_{s,k})$. Recall that ${}_yB_{s,k}$ is the ball of radius $y\sigma$ in $X_{s,e}$ which has the same center point as does $B_{s,k}$. The following notation will prove convenient:

$$
\begin{aligned}
{}_k\hat{Y}_j(f) &= g_e({}_kY_j(f) \times {}_5B_{s,k}); \\
{}_k\hat{Y}_j'(f) &= g_e({}_kY_j'(f) \times {}_5B_{s,k}); \\
{}_k\hat{Y}_j(f;t) &= t\text{'th level thickening of } {}_k\hat{Y}_j(f).
\end{aligned}
$$

Recall that the set of balls $H(k,j,f;k')$ has been defined for each k' in 9.1. We let $\hat{H}(k,j,f;k')$ denote all the ${}_{k'}\hat{Y}_{j'}'(f')$ such that ${}_{k'}Y_{j'}(f')$ is a member of $H(k,j,f;k')$. Now define ${}_k\hat{Y}_j(f;1)$ to be the union of ${}_k\hat{Y}_j'(f)$ with all sets of the form

$$F^{-q}({}_{k'}\hat{Y}_{j'}'(f')) \cap g_e(X_{u,e} \times {}_5B_{s,k}),$$

where $k \in I_e$ and $f \in K_{u,e}$ are both fixed, and $k' \in I_{e'}$, $f' \in K_{u,e'}$, $e' \in C$ are all variables, and ${}_{k'}\hat{Y}_{j'}(f')$ must be a member of $\hat{H}(k,j,f;k')$. The higher thickenings are defined by induction. Suppose that all the t'th level thickenings have been defined, and for each k' let $H^t(k,j,f;k')$ denote all the sets ${}_{k'}\hat{Y}_{j'}(f';t)$ such that ${}_{k'}\hat{Y}_{j'}(f')$ is a member of $\hat{H}(k,j,f;k')$. Then define ${}_k\hat{Y}_j(f;t+1)$ to be the union of ${}_k\hat{Y}_j'(f)$ with all sets of the form

$$F^{-q}({}_{k'}\hat{Y}_{j'}(f';t)) \cap g_e(X_{u,e} \times {}_5B_{s,k}),$$

where ${}_{k'}\hat{Y}_{j'}(f';t)$ must be a member of $H^t(k,j,f;k')$. Finally we can define ${}_k\hat{Y}_j(f;\infty)$ by

$${}_k\hat{Y}_j(f;\infty) = \overline{\bigcup_{t \geq 1} {}_k\hat{Y}_j(f;t)}.$$

We now state in the following one theorem a reformulation of both theorems 9.6 and 12.1 that is true even when hypothesis 5.2 is not assumed to hold.

THEOREM 14.19. *For each $e \in C$, $k \in I_e$, $t \in \{1,2,\ldots,\infty\}$ there is an embedding $h_{k,t} : g_e({}_{n+1}N_{u,e} \times {}_2B_{s,k}) \to g_e(X_{u,e} \times {}_2B_{s,k})$ satisfying the following property. Suppose that for some ball ${}_{k'}Y_{i'}(f)$ we have that $g_{e'}({}_{k'}Y_{i'}(f) \times {}_2B_{s,k'})$ intersects with $g_e({}_0N_{u,e} \times {}_2B_{s,k})$—where $e' \in C$, $k' \in I_{e'}$, and $f \in K_{u,e'}$. Also suppose that ${}_0N_{s,e} \cap {}_2B_{s,k} \neq \emptyset$. Then we must have that*

$$h_{k,t}({}_{k'}\hat{Y}_{i'}(f) \cap g_e({}_{n+1}N_{u,e} \times {}_2B_{s,k})) = {}_{k'}\hat{Y}_{i'}(f;t) \cap g_e({}_{n+1}N_{u,e} \times {}_2B_{s,k}).$$

PROOF OF THEOREM 14.19. The proof consists of the following four steps.

STEP I. In this step we construct an embedding $h_{k,t} : g_e({}_{n+1}N_{u,e} \times b) \to g_e(X_{u,e} \times b)$ which satisfies the following property, where b denotes the center of $B_{s,k}$.

<u>14.20.</u> Let ${}_{k'}Y_{i'}(f)$ be any ball as in 14.19. Then we must have that

$$h_{k,t}({}_{k'}\hat{Y}_{i'}(f) \cap g_e({}_{n+1}N_{u,e} \times b)) = {}_{k'}\hat{Y}_{i'}(f;t) \cap g_e({}_{n+1}N_{u,e} \times b).$$

The construction of $h_{k,t} : g_e({}_{n+1}N_{u,e} \times b) \to g_e(X_{u,e} \times b)$ consists of a modulo $(\varepsilon,\varepsilon')$ version—in the sense of 14.12—of the proof of the thickening theorem 9.6 (if $t < \infty$) or of a modulo $(\varepsilon,\varepsilon')$ version of the proof of the limit theorem 12.1 (if $t = \infty$). We will give a few more details in the case that $t < \infty$. Define $\check{B}_{s,k}$ as in 11.0, and for each positive integer j define ${}_jX_{u,e}$ as in 11.1. The definition of the triangulation $L_{i,j}$ given in 11.1(b) must be slightly modified as follows.

<u>14.21.</u> A smooth simplex $\Delta \subset {}_jX_{u,e}$ is in $L_{i,j}$ if and only if there is $e' \in C$, $\Delta' \in K_{u,e'}$, and $k' \in I_{e'}$ so that the following hold.

(a) $\dim(e') = i$.

(b) $F^{qj} \circ g_e(\Delta \times \check{B}_{s,k}) \cap g_{e'}(\Delta' \times B_{s,k'}) \neq \emptyset$.

(c) $F^{qj} \circ g_e(\Delta \times b) = g_{e'}(\Delta' \times {}_5B_{s,k'}) \cap F^{qj} \circ g_e(Y \times b)$, for some neighborhood Y for Δ in B^n_δ.

Now the smooth triangulations $K_{i,j}$ can be defined as in 11.1(c)(d), with the slight modification that we only require 11.1(d) to be true modulo $(\varepsilon,\varepsilon')$ in the sense of 14.12. Of course now the $K_{i,j}$ only satisfy modulo $(\varepsilon,\varepsilon')$-versions of 11.2(a)–(c).

For each $i \in \{0,1,\ldots,n+m\}$ and each $\Delta \in K_{i,j}$ we define in 11.3 P.S. balls $\{Y_\ell(\Delta) : 1 \leq \ell \leq \phi(\Delta)\}$. This definition must be modified as follows.

14.22. For each $Y_\ell(\Delta)$ there must be $e' \in C$, $k' \in I_{e'}$, $\Delta' \in K_{u,e'}$, and a ball ${}_{k'}Y_{\ell'}(\Delta')$—as in 8.2—so that the following hold.

(a) Properties 14.21(b)(c) both hold.

(b) $F^{qj} \circ g_e(Y_\ell(\Delta) \times b) = g_{e'}({}_{k'}Y_{\ell'}(\Delta') \times {}_5B_{s,k'}) \cap F^{qj} \circ g_e(Y \times b)$, for some neighborhood Y for $\Delta \cup Y_\ell(\Delta)$.

Note that the $\{K_{i,j}\}$ and the balls $\{Y_\ell(\Delta)\}$ satisfy modulo $(\varepsilon, \varepsilon')$ versions of both 11.4 and 11.5 (in the sense of 14.12).

Now we carry out modulo $(\varepsilon, \varepsilon')$ versions of each of the three steps preceding 11.5 to complete the proof of 14.20 in the case that $t < \infty$.

In the preceding paragraphs we have constantly referred to "modulo $(\varepsilon, \varepsilon')$ versions" of arguments and properties without being specific about the values of ε and ε'. We shall correct that now with the stipulations:

14.23.

$$\varepsilon = \gamma_4 r(\delta)\delta,$$
$$\varepsilon' = \gamma_4 r(\delta),$$

where $r(\ )$, δ come from 14.1 and 14.2, and γ_4 is the maximum of γ_2 and γ_3. (See 14.11(b) for γ_2 and see 14.13, 14.16 for γ_3.) We note that 14.23 gives the $(\varepsilon, \varepsilon')$, but we do not always use the standard metric on $X_{u,e}$ to discuss the concept of "modulo $(\varepsilon, \varepsilon')$" in the preceding paragraphs. In fact for each positive integer j we let $\langle , \rangle_j$ denote the Riemannian metric on ${}_jX_{u,e}$ which is gotten by pulling back the metric on M along the map $F^{qj} \circ g_e : {}_jX_{u,e} \times b \to M$. Then the modulo $(\varepsilon, \varepsilon')$ versions of properties 11.2, 11.4, 11.5 hold for the $\{K_{i,j} : 0 \le i \le n+m\}$ and the $\{Y_\ell(\Delta) : \Delta \in K_{i,j},\ 0 \le i \le n+m, \text{ and } 1 \le \ell \le \phi(\Delta)\}$ for ε and ε' as in 14.23 only *when all metric measurements are carried out with respect to the metric* $\langle , \rangle_j$.

This completes step I.

STEP II. In this step we construct for each positive integer t a smooth fiber bundle projection $P_t : g_e({}_{n+1}N_{u,e} \times {}_2B_{s,k}) \to g_e({}_{n+1}N_{u,e} \times b)$ which satisfies the following property.

14.24. For any ball ${}_{k'}Y_{i'}(f)$ as in 14.19 we must have that

$$P_t({}_{k'}\hat{Y}_{i'}(f;t) \cap g_e({}_{n+1}N_{u,e} \times {}_2B_{s,k})) = {}_{k'}\hat{Y}_{i'}(f;t) \cap g_e({}_{n+1}N_{u,e} \times b).$$

For each non-negative integer j let $B_j(,)$ denote the Riemannian metric on $g_e({}_jX_{u,e} \times {}_2B_{s,k})$ gotten by pulling the Riemannian metric on M back along the map $F^{qj} : g_e({}_jX_{u,e} \times {}_2B_{s,k}) \to M$. Note that it follows from 14.2(c) and a tapering argument that there is a smooth bundle projection $\rho_j : g_e({}_jX_{u,e} \times {}_2B_{s,k}) \to g_e({}_jX_{u,e} \times b)$ which satisfies the following properties.

14.25. For each $e' \in C$ there is a smooth map $f_{e'} : A_{e'} \to g_e({}_jX_{u,e} \times b)$ defined as follows. Let $A_{e'}$ denote the interior of $B_{e'}$, where $B_{e'}$ is the union of all connected subsets Y of the intersection $F^{-qj}(g_{e'}(p \times X_{s,e'})) \cap g_e({}_jX_{u,e} \times {}_2B_{s,k})$ which satisfy the following: $p \in X_{u,e'}$; $Y \cap g_e({}_jX_{u,e} \times b)$ is a single point (this is the same as requiring that $Y \cap g_e({}_jX_{u,e} \times b)$ is non-empty). Each $x \in A_{e'}$ lies in a unique such Y—denoted by Y_x. Define $f_{e'}(x)$ to be equal to the point $Y_x \cap g_e({}_jX_{u,e} \times b)$. Then the following must be true.

(a) There is a number $\gamma_5 > 0$ which depends only on $\dim(M)$.

(b) For any $e' \in C$ the C^0-distance from $f_{e'}$ to $\rho_j|A_{e'}$—with respect to the metric $B_j(,)$—is less than $\gamma_5 r(\delta)\delta$. The C^1-distance from $f_{e'}$ to $\rho_j|A_{e'}$—with respect to the metric $B_j(,)$—is less than $\gamma_5 r(\delta)$. Here $r(\ )$, and δ come from 14.1 and 14.2.

Note that it follows from 14.2, 14.8, 14.9, 14.11, and 14.13 that there is a smooth bundle projection $\overline{\rho}_j : g_e({}_jX_{u,e} \times {}_2B_{s,k}) \to g_e({}_jX_{u,e} \times b)$ which satisfies the following properties.

14.26. (a) For each ${}_{k'}Y_{\ell'}(\Delta')$ and $Y_\ell(\Delta)$ as in 14.22 let ${}_{k'}Z_{\ell'}(\Delta')$ denote the connected component of $F^{-qj} \circ g_{e'}({}_{k'}Y_{\ell'}(\Delta') \times {}_5B_{s,k}) \cap g_e({}_jX_{u,e} \times {}_2B_{s,k})$ which contains $Y_\ell(\Delta) \times b$. Then we must have $\overline{\rho}_j({}_{k'}Z_{\ell'}(\Delta')) = g_e(Y_\ell(\Delta) \times b)$.

(b) The C^0-distance from ρ_j to $\overline{\rho}_j$—measured with respect to the metric $B_j(,)$—is less than $\gamma_6 r(\delta)\delta$. The C^1-distance from ρ_j to $\overline{\rho}_j$—measured with respect to the metric $B_j(,)$—is less than $\gamma_6 r(\delta)$. Here $r(\ )$, δ come from 14.1, 14.2, and $\gamma_6 > 0$ depends only on $\dim(M)$ and on r_{n+m} of 14.11.

Now the idea is to construct the projection $P_t : g_e({}_{n+1}N_{u,e} \times {}_2B_{s,k}) \to g_e({}_{n+1}N_{u,e} \times b)$ from the $\overline{\rho}_0, \overline{\rho}_1, \overline{\rho}_2, \ldots, \overline{\rho}_t$ by a tapering procedure so that on a large portion of the set

$$g_e(({}_{n+1}N_{u,e} \cap {}_jX_{u,e}) \times {}_2B_{s,k}) - g_e({}_{j+1}X_{u,e} \times {}_2B_{s,k})$$

we have that $P_t = \overline{\rho}_j$. To be more specific we require that the P_t satisfy the following properties in addition to 14.24.

14.27. (a) $P_t|D_j = \overline{\rho}_j|D_j$, where D_j is a large portion of the set

$$g_e(({}_{n+1}N_{u,e} \cap {}_jX_{u,e}) \times {}_2B_{s,k}) - g_e({}_{j+1}X_{u,e} \times {}_2B_{s,k}).$$

(b) There is a number $\gamma_7 > 0$ which depends only on $\dim(M)$ and on r_{n+m} of 14.11 and on q. Set

$$E_j = g_e(({}_{n+1}N_{u,e} \cap {}_jX_{u,e}) \times {}_2B_{s,k}) - g_e({}_{j+2}X_{u,e} \times {}_2B_{s,k}).$$

(c) The C^0-distance from $P_t|E_j$ to $\overline{p}_j|E_j$—with respect to the metric $B_j(,)$—is less than $\gamma_7 r(\delta)\delta$. The C^1-distance from $P_t|E_j$ to $\overline{p}_j|E_j$—with respect to the metric $B_j(,)$—is less than $\gamma_7 r(\delta)$.

(d) $P_t \left| \bigcup_{i=0}^{j} D_i = P_j \right| \bigcup_{i=0}^{j} D_i$, for each $0 \leq j \leq t$.

The remaining details in this step are left to the reader. (Note that 14.3(b) will be of use in verifying that 14.27 is possible.)

This completes step II.

STEP III. In this step we construct a continuous fiber bundle projection $P_\infty : g_e({}_{n+1}N_{u,e} \times {}_2B_{s,k}) \to g_e({}_{n+1}N_{u,e} \times b)$ which satisfies the following property.

14.28. For any ${}_{k'}Y_{i'}(f)$ as in 14.19 we must have that

$$P_\infty({}_{k'}\hat{Y}_{i'}(f;\infty) \cap g_e({}_{n+1}N_{u,e} \times {}_2B_{s,k}) = {}_{k'}\hat{Y}_{i'}(f;\infty) \cap g_e({}_{n+1}N_{u,e} \times b).$$

We define $P_\infty = \lim_{t\to\infty} P_t$, where the limit is taken with respect to the metric $B_0(,)$ on $g_e({}_{n+1}N_{u,e} \times {}_2B_{s,k})$. Note that 14.25–14.27, 5.1, and 14.3(b) together imply that this limit exists and is a continuous fiber bundle projection. The details of this verification are left to the reader whom we leave with the following hint. Note that the fibers of P_∞ are of two types: if $p \in g_e({}_{n+1}N_{u,e} \times b) - g_e({}_{j+1}X_{u,e} \times b)$ for some non-negative integer j then the fiber $P_\infty^{-1}(p)$ is equal to $P_{j+1}^{-1}(p)$; if $p \in g_e({}_{n+1}N_{u,e} \times b) \bigcap \left(\bigcap_{j\geq 0} g_e({}_jX_{u,e} \times b) \right)$ then the fiber $P_\infty^{-1}(p)$ is a smooth manifold contained in some leaf of the stable foliation $W_s(\Lambda)$.

It follows from the definition of P_∞, from 14.24, and from the definition of the ${}_{k'}\hat{Y}_{i'}(f;\infty)$ as $\lim_{t\to\infty} {}_{k'}\hat{Y}_{i'}(f;t)$ that P_∞ satisfies 14.28.

STEP IV. We can complete the proof of theorem 14.19 by simply combining the results of steps I, II, and III (i.e., combining 14.20, 14.24, and 14.28).

We begin by noting that each of the fiber bundles

14.29. $P_t : g_e({}_{n+1}N_{u,e} \times {}_2B_{s,k}) \to g_e({}_{n+1}N_{u,e} \times b)$, $t \in \{0, 1, \ldots, \infty\}$, is a trivial fiber bundle. One way of seeing this is to note that there will be no loss of generality in assuming that there is a subset $C_{u,e} \subset B_\delta^n$ satisfying the following.

14.30. (a) $C_{u,e}$ is contractible; $X_{u,e} \subset C_{u,e}$.

(b) The construction of the P_t given in steps II and III may be applied to the larger set $C_{u,e} \times {}_2B_{s,k}$ to get an extension of the fiber bundles of 14.29 to a fiber bundle

$$P_t : g_e(C_{u,e} \times {}_2B_{s,k}) \to g_e(C_{u,e} \times b).$$

Note that 14.30 makes possible the extension of the embedding $h_{k,t} : g_e({}_{n+1}N_{u,e} \times b) \to g_e(X_{u,e} \times b)$ of 14.20 to an embedding $h_{k,t} : g_e({}_{n+1}N_{u,e} \times {}_2B_{s,k}) \to g_e(X_{u,e} \times {}_2B_{s,k})$ which maps the fibers of the bundle projection P_0 onto the fibers of the bundle projection P_t. Now it follows from 14.20, 14.24, and 14.28 that the embedding $h_{k,t} : g_e({}_{n+1}N_{u,e} \times {}_2B_{s,k}) \to g_e(X_{u,e} \times {}_2B_{s,k})$ satisfies the conclusions of theorem 14.19.

This completes the proof of theorem 14.19.

We come now to the last modification which must be made to the proof of theorem 1.5 (when 5.2 is satisfied) in order to get a proof that works when 5.2 is not satisfied. This modification occurs in the construction of the Markov cell structure given in section 13. Since the notation ${}_k\hat{Y}_i(f;\infty)$ has also been used in section 13, but with a different meaning than that attached to it in this section, we shall defer to the earlier meaning of this notation given in section 13. *From now on let us use the notation* ${}_kY_i'(f;\infty)$ *to symbolize what has to this point in this section always been denoted by* ${}_k\hat{Y}_i(f;\infty)$, *and let us use* ${}_k\hat{Y}_i(f;\infty)$ *to denote the meaning of this symbol discussed just prior to 13.1.* Here are some of the modifications which must be made in section 13.

14.31. (a) Replace $g_e({}_kY_i(f;\infty) \times_\ell \hat{Y}_j(g;\infty))$—where $f \in K_{u,e}$, $g \in \hat{K}_{u,e}$—everywhere by the intersection ${}_kY_i'(f;\infty) \cap {}_\ell\hat{Y}_j'(g;\infty)$, e.g. in 13.3, 13.6, etc. Here the ${}_\ell\hat{Y}_j'(g;\infty)$ are constructed just as the ${}_kY_i'(f;\infty)$ have been constructed except the roles of the factors of $B_\delta^n \times B_\delta^m$ are reversed. (Recall that $B_\delta^n \times B_\delta^m$ is the domain of the maps g_i of 14.4.)

(b) In 13.6 (and in the paragraph preceding 13.6) replace the requirements $A_{u,e} \cap {}_j\hat{N}_{s,e}' \neq \emptyset$ and $A_{s,e} \cap {}_jN_{s,e}' \neq \emptyset$ by the single requirement $A \cap g_e({}_j\hat{N}_{s,e}' \times {}_jN_{s,e}') \neq \emptyset$.

(c) We must redefine the partition $A(x)$ of 13.9. Note that each set B in $J(A,x)$—see 13.6 and (b) above for $J(A,x)$—has the form ${}_{k'}Y_{i'}'(f';\infty) \cap {}_{\ell'}\hat{Y}_{j'}'(g';\infty)$. Let $B_1, B_2, \dots, B_t$ denote all the ${}_{k'}Y_{i'}'(f';\infty)$ which occur for all the sets B in $J(A,x)$, and let $\hat{B}_1, \hat{B}_2, \dots, \hat{B}_t$ be all the ${}_{\ell'}\hat{Y}_{j'}'(g;\infty)$ which occur for all the sets B in $J(A,x)$. Define $A(x)$ to be the partition of A generated under the operation of set intersection by all the sets $\{B_i \cap A,\ \partial B_i \cap A,\ \overline{A - B_i} : 1 \leq i \leq t\}$ and all the sets $\{\hat{B}_i \cap A,\ \partial\hat{B}_i \cap A,\ \overline{A - \hat{B}_i} : 1 \leq i \leq t\}$.

Now define the Markov cell structure as in 13.11. The remaining details in checking that we do indeed have a Markov cell structure for $F^q : M \to M$ near Λ are left to the reader.

§15. SELECTED PROBLEMS

In this section we formulate some new problems and update some old problems dealing with hyperbolic sets of a diffeomorphism. We also give two applications of Markov cell structures (see 15.4, 15.8). Our problems are organized under the following four headings.

I. LOCAL TOPOLOGICAL TYPES OF HYPERBOLIC SUBSETS.

Let $F : M \to M$ denote a diffeomorphism of the smooth manifold M and let $\Lambda \subset M$ be a hyperbolic set for $F : M \to M$. A natural problem is to characterize the local topological type of the embedding $\Lambda \subset M$. A weaker form of this problem was first stated by S. Smale [26] who asked if every hyperbolic set must be either a manifold or locally homeomorphic to Euclidean space crossed with a zero-dimensional space. A negative answer to Smale's question was first given by J. Guckenheimer [10]; and hyperbolic sets having a wide variety of local homotopy types were later constructed by L. Jones [13]. Further examples have recently been constructed by H. Bothe [3]. Despite all of these examples there are the following special cases of Smale's original problem which are still unanswered.

PROBLEM 15.1. (a) Suppose that $\Lambda \subset M$ is hyperbolic set for $F : M \to M$ which is an attractor. Then is Λ locally homeomorphic to Euclidean space crossed with a zero dimensional space?

(b) Suppose that $\Lambda \subset M$ is a hyperbolic set for $F : M \to M$ which is locally connected. Then is Λ a manifold?

We can apply Markov cell structures to get a partial classification of the local topological types of embeddings $\Lambda \subset M$ of hyperbolic subsets. Roughly speaking we generalize the "middle third" construction (which yields the Cantor set as a subset of the interval $I = [0, 1]$) to obtain a class of compact subsets of the cubes I^m, $m \geq 1$, which we call "recurrent subsets." The partial classification then states that the embedding $\Lambda \subset M$ is locally homeomorphic to a recurrent subset in I^m.

We shall now give a precise description of these results. The following definition is needed to generalize the middle third construction of the Cantor set. Recall that a regular cell structure (or regular cell complex) is a cell complex C such that for each cell $e \in C$ the boundary of e is embedded in C as a subcomplex.

DEFINITION 15.2. *A* **recurrent cell structure** *in I^m consists of a sequence of finite regular cell complexes $C_1, C_2, \ldots$ in I^m, a pair of finite* *index sets J, J' with $J' \subset J$, and mappings $f_k : C_k \to J$, $k = 1, 2, \ldots$, all satisfying the following properties.*

(a) *The underlying set of C_1, denoted $|C_1|$, is equal I^m.*

(b) *For each $k \geq 1$ there are the following relations between C_k and C_{k+1}:*

(i) $|C_{k+1}| \subset |C_k|$;

(ii) *The collection of cells $f_k^{-1}(J - J')$ is a subcomplex of C_k, denoted by C_k', and C_{k+1} is a subdivision of C_k'.*

(c) $\lim_{k\to\infty} D(C_k) = 0$, *where $D(C_k) =$ maximum $\{\text{diameter}(e) : e \in C_k\}$.*

(d) *For each $k \geq 1$ and each $e \in C_k$ let $C_k(e)$ denote the subcomplex of all $d \in C_k$ satisfying $d \subset e$. The restriction maps $f_k|C_k(e)$ satisfy the following:*

(i) *Each $f_k|C_k(e)$ is an embedding into J.*

(ii) *Suppose for any $k, k' \geq 1$ and any $e \in C_k$, $e' \in C_{k'}$, we have that $f_k(C_k(e)) = f_{k'}(C_{k'}(e'))$, then there is a homeomorphism $h : e \to e'$ which maps the cells of $C_k(e)$ onto the cells of $C_{k'}(e')$ and satisfies $f_k(d) = f_{k'}(h(d))$ for all $d \in C_k(e)$.*

(iii) *Given $k > 1$ and any $e \in C_k$ there is a $d \in C_1$ such that $f_1(C_1(d)) = f_k(C_k(e))$.*

Now we can define a recurrent subset of the m-cube I^m.

DEFINITION 15.3. *A subset $X \subset I^m$ is called a* **recurrent subset** *if there is a recurrent cell structure in I^m (that is $f_k : C_k \to J$, $k = 1, 2, \ldots, J'$, as in 15.2) such that $X = \cap_{k=1}^{\infty} |C_k|$. We call the $\{f_k : C_k \to J\}$, $J' \subset J$, a recurrent cell structure for X.*

EXAMPLE. The "middle third" construction for the Cantor subset $X \subset I$ can be seen to be a recurrent cell structure for X as follows. Set C_1 equal the interval $[0, 1]$ having the end points 1,0 for vertices. Let $|C_k|$, $k \geq 1$, be the result of applying the middle third construction to [0,1] $k - 1$ times (C_k is the minimal cell structure for $|C_k|$). Set $J = \{1, 2, 3, 1^+, 1^-, 2^+, 2^-, 3^+, 3^-\}$, and set $J' = \{2\}$. $f_k : C_k^1 \to J$ maps the first, middle, and last third of each edge of C_{k-1} to 1,2,3 respectively; $f_k : C_k^0 \to J$ maps the left and right end points of each edge $e \in C_k^1$ to i^- and i^+ respectively where $i = f_k(e)$.

The proof of the following theorem, which is just an application of Markov cell structures, will be given at the end of this section.

THEOREM 15.4. *Let $\Lambda \subset M$ be a hyperbolic set for the diffeomorphism $F : M \to M$. Then for each point $p \in \Lambda$ there is a neighborhood $U \subset M$*

for p in M such that the pair $(U, U \cap \Lambda)$ is homeomorphic to a pair (I^m, X) where $X \subset I^m$ is a recurrent subset of the m-cube I^m.

II. Finite representations for dynamical systems.

R. Bowen [5] has shown that for any Axiom A diffeomorphism $F : M \to M$ on a compact closed manifold M the restriction of F to its non-wandering set $\Omega \subset M$ has a finite presentation. In fact Bowen shows that each Markov partition for $F : \Omega \to \Omega$ gives rise to a finite presentation. The complementary part of F, i.e., $F : M - \Omega \to M - \Omega$, behaves like a covering transformation, so it is finitely presented on arbitrarily large compact subsets of $M - \Omega$. These observations lead us to the next problem.

PROBLEM 15.5. (a) Suppose that $F : M \to M$ is an Axiom A diffeomorphism on the compact closed manifold M. Then does $F : M \to M$ have a finite presentation?

(b) Same question as part (a) when $F : M \to M$ is assumed to be a structurally stable diffeomorphism.

(c) Same question as part (a) when $F : M \to M$ is assumed to be an Axiom A diffeomorphism satisfying the strong transversality condition.

Markov cell structures may contribute to the solution of 15.5 in the following way. Let us return to the situation where $F : M \to M$ is a diffeomorphism of the smooth manifold M and $\Lambda \subset M$ is a hyperbolic set for $F : M \to M$. Bowen's results [5] can be extended to get a finite presentation for $F : \Lambda \to \Lambda$. It seems likely that a Markov cell structure for $F : M \to M$ near Λ will give rise to a finite presentation for some power of $F : M \to M$ near Λ. We shall now give a precise formulation of this conjecture.

DEFINITION 15.6. *Let C denote a Markov cell structure for $F : M \to M$ near the hyperbolic set $\Lambda \subset M$. The* **Markov pairings** *associated to C and F are the maps $\phi_u, \phi_s, \psi : C \times C \to Z_2$ defined as follows. For $e_1, e_2 \in C$ set $\phi_u(e_1, e_2) = 1$ if $e_1 \subset \partial_u e_2$ and set $\phi_u(e_1, e_2) = 0$ otherwise; set $\phi_s(e_1, e_2) = 1$ if $e_1 \subset \partial_s e_2$ and set $\phi_s(e_1, e_2) = 0$ otherwise; set $\psi(e_1, e_2) = 1$ if $(e_1 - \partial e_1) \cap F^q(e_2 - \partial e_2) \neq \emptyset$ and set $\psi(e_1, e_2) = 0$ otherwise.*

CONJECTURE 15.7. Let C denote a Markov cell structure for $F : M \to M$ near the hyperbolic set $\Lambda \subset M$ (with respect to the integer q of 1.3). Let $\phi_u, \phi_s : C \times C \to Z_2$, $\psi : C \times C \to Z_2$ denote the Markov pairings. Then there is a neighborhood V for Λ in $|C|$ such that the map $F^q : V \to |C|$ can be reconstructed from the Markov pairings ϕ_u, ϕ_s, ψ. That is, beginning only with the finite amount of information in the pairings ϕ_u, ϕ_s, ψ, one can construct a pair of topological spaces $X \subset Y$

and an embedding $g : X \to Y$ and prove that there is a homeomorphism $h : (Y, X) \to (|C|, V)$ such that $h \circ g(x) = F^q \circ h(x)$ holds for all $x \in X$. This conjecture can be rephrased as follows: *the Markov pairings $\phi_u, \phi_s, \psi : C \times C \to Z_2$ are a finite presentation for $F^q : M \to M$ near Λ.*

The proof of the following special case of conjecture 15.7 can be found at the end of this section.

THEOREM 15.8. *Suppose that $\Lambda \subset M$ is a hyperbolic set for $F : M \to M$ which is also an attractor. Let C denote a Markov cell structure for $F : M \to M$ near Λ (with respect to the integer q of 1.3). Then the Markov pairings $\phi_u, \phi_s : C \times C \to Z_2$ and $\psi : C \times C \to Z_2$ are a finite presentation for $F^q : M \to M$ near $\Lambda \subset M$.*

In view of 15.7 and 15.8 it would seem that the following problem would be a prudent first step towards a solution to problem 15.5(c).

PROBLEM 15.9. Let $F : M \to M$ be an Axiom A diffeomorphism satisfying the strong transversality condition (on the compact closed manifold M). Formulate the meaning of and prove that existence of "Markov cell structures" for $F : M \to M$ on *all of M*.

III. NEW TOPOLOGICAL INVARIANTS?

Can Markov cell structures for a diffeomorphism $F : M \to M$ near one of its hyperbolic sets $\Lambda \subset M$ be used to define new topological invariants for $F : M \to M$ near Λ? One might draw an analogy with homology theory: by triangulating the manifold M one arrives at the first description of the integral homology groups as chains of triangles. Markov cell structures are a type of "triangulation theory" for diffeomorphisms near their hyperbolic sets: is there a "homology theory" associated to this "triangulation theory"? Of course triangulations of M are not needed to discuss the integral homology of M so this analogy may be misleading. However cell structures are needed to discuss the torsion invariants of Whitehead and Reidermeister (cf. [17]). Do Markov cell structures give rise to new "torsion" type invariants for $F : M \to M$?

If new topological invariants are to arise from Markov cell structures by analogies with known topological invariants, as suggested in the preceding paragraph, then we believe a first step towards discovering these invariants would be to address the following problem.

PROBLEM 15.10. Let C_1, C_2 be two Markov cell structures for $F : M \to M$ near the hyperbolic set $\Lambda \subset M$. Formulate what it means for C_1, C_2 to be "equivalent up to subdivision," and prove that if C_1, C_2 are obtained by the constructions of sections 2–14 in this paper then they are equivalent up to subdivision.

Can recurrent cell structures for recurrent subsets $X \subset I^m$ be used to define new topological invariants for the topological pair (I^m, X)? Here also it would seem that the first step towards discovering these new invariants would be to address the problem of when two recurrent cell structures are "equivalent up to subdivision."

IV. VANISHING PONTRJAGIN CLASSES?

Let $\Lambda \subset M$ be a hyperbolic set for the diffeomorphism $F : M \to M$ and let ξ_u, ξ_s be the linear bundles over Λ given in 1.1. Choose classifying maps $g_u : \Lambda \to B0$, $g_s : \Lambda \to B0$ for these bundles. (Note that proposition 5.1 assures the existence of g_u, g_s.) Let $P^{4i} \in H^{4i}(B0, Q)$ denote the rational Pontrjagin class of dimension $4i$ for the universal bundle over the classifying space $B0$. Note that P^{4i} may be thought of as either a singular cohomology class—in which case we use the notation P_s^{4i}—or as a Cech cohomology class—in which case we use the notation P_c^{4i}—because the singular and Cech cohomologies of the space $B0$ are identical. Define the **singular rational Pontrjagin class** in dimension $4i$ for the bundles ξ_u, ξ_s—denoted by $P_s^{4i}(\xi_u)$, $P_s^{4i}(\xi_s)$—to be the images of P_s^{4i} under the maps induced by g_u, g_s on singular cohomology theory with rational coefficients. Define the **Cech rational Pontrjagin class** in dimension $4i$ for the bundles ξ_u, ξ_s—denoted by $P_c^{4i}(\xi_u), P_c^{4i}(\xi_s)$—to be the image of P_c^{4i} under the maps induced by g_u, g_s on Cech cohomology theory with rational coefficients.

CONJECTURE 15.11. For each integer $i > 0$ we must have $P_s^{4i}(\xi_u) = 0$ and $P_s^{4i}(\xi_s) = 0$. In particular the *singular* rational Pontrjagin classes for $T(M)|\Lambda$ must vanish above dimension zero for any hyperbolic set $\Lambda \subset M$ of a diffeomorphism $F : M \to M$.

REMARK. In support of this conjecture we have the following evidence. All known examples of hyperbolic sets $\Lambda \subset M$ for diffeomorphisms $F : M \to M$ are of the following four types:

(a) $F : \Lambda \to \Lambda$ is an Anosov diffeomorphism and both of the bundles ξ_u, ξ_s are flat linear bundles with finite structure groups.

(b) $F : \Lambda \to \Lambda$ is an expanding attractor (cf. [28]).

(c) $F : \Lambda \to \Lambda$ is one of the hyperbolic sets constructed in [13].

(d) Products of the types in (a), (b), or (c).

If the singular rational Pontrjagin classes for types (a), (b), and (c) vanish then so do those for type (d) (apply the product formula for characteristic classes). In case (a) it is known that the rational Pontrjagin classes of a flat linear bundle having finite structure group must vanish in dimensions above zero (cf. [16]). In case (b) it can be shown that each path component of Λ is homeomorphic to a Euclidean space

R^n. In case (c) it can be shown that each path component X of Λ can be embedded in a Euclidean space, $X \subset R^n$, in such a way that the restricted bundles $\xi_u|X$ and $\xi_s|X$ extend to linear bundles ξ'_u, ξ'_s on all of R^m. Thus in both cases (b), (c) the singular rational Pontrjagin classes for ξ_u and ξ_s must vanish in dimensions above zero.

REMARK. Conjecture 15.11 is false when applied to the Cech rational Pontrjagin classes. To see this we construct a hyperbolic set $F : \Lambda \to \Lambda$ which is an expanding attractor such that $P^4_c(\xi_u) \neq 0$. We use theorem 8.1 in [8] to choose a regular cell structure C for a smooth manifold N such that the codimension 1 skeleton C^{n-1} of C is a branched manifold supporting an expanding immersion $I : C^{n-1} \to C^{n-1}$ such that $I : C^{n-1} \to N$ is homotopic to the inclusion map $C^{n-1} \subset N$. We also assume that $\dim(N) \geq 6$ and that $P^4(T(N)) \neq 0$. Now use the theory of R.F. Williams [28] to construct from $I : C^{n-1} \to C^{n-1}$ the desired expanding attractor $F : \Lambda \to \Lambda$. We note that Williams' theory only provides us with a one-one immersion $F : V \to M$, which extends $F : \Lambda \to \Lambda$ to an open neighborhood V for Λ in some smooth manifold M, such that $F : V \to M$ satisfies the properties of 1.1. It seems likely that in many cases we should be able to extend $F : V \to M$ to a diffeomorphism $F : M \to M$.

PROOF OF THEOREM 15.4.

Choose a Markov cell structure C for $F : M \to M$ near the hyperbolic set $\Lambda \subset M$. We claim that for any $p \in \Lambda$ the Markov cell structure C can be chosen to satisfy in addition to 1.3(a)(b) the following property.

15.12. There is an m-dimensional cell $e \in C$ ($m = \dim(M)$) such that $p \in \mathrm{Int}(e)$.

To verify 15.12 one must be familiar with the constructions in sections 2–14 above. The details are left to the reader.

In theorem 15.4 let U denote e of 15.12. Note that U is homeomorphic to I^m; this makes $\Lambda \cap U$ a subset $X \subset I^m$. It will suffice to show that $X \subset I^m$ is a recurrent subset of I^m having a recurrent cell structure $f_k : C_k \to J$, $k = 1, 2, 3, \ldots$, $J' \subset J$, which satisfies 15.2(a)–(d). Define J to be the Cartesian product $C \times C$. Define $J' \subset J$ to be the collection of all pairs of cells $(e, d) \in C \times C$ such that one of the following properties holds: either $F^q(e) \cap M - |C| \neq \emptyset$; or $F^{-q}(d) \cap M - |C| \neq \emptyset$. Each cell $\Delta \in C_k$ has the following form:

15.13. (a) Δ is the closure of a connected component of $U \cap e \cap F^{-kq}(e_1 - \partial e_1) \cap F^{kq}(e_2 - \partial e_2)$. Here $e_1, e_2 \in C$, and $e \in C_{k-1}$ satisfies $F^{iq}(e) \subset |C|$ for all $i = 0, 1, -1, 2, -2, \ldots, k, -k$. (If $k = 1$ then set $C_{k-1} = C$.)

For $\Delta \in C_k$ as in 15.13(a) define $f_k(\Delta)$ as follows.

15.13. (b) $f_k(\Delta) = (e_1, e_2)$, where e_1, e_2 come from 15.13(a).

The sequence $f_k : C_k \to J$, $k = 1, 2, 3, \ldots$, satisfies all of 5.2(a)–(d) with the possible exception of 5.2(d)(iii). To get 5.2(d)(iii) satisfied also we may have to replace the sequence $f_k : C_k \to J$, $k = 1, 2, 3, \ldots$, by the sequence $f_k : C_k \to J$, $k = j, j+1, j+2, \ldots$, for sufficiently large j. The remaining details are left to the reader.

This completes the proof of theorem 15.4.

PROOF OF THEOREM 15.8. If ε is chosen sufficiently small in 1.5 then we may assume (after replacing C by a subcomplex if necessary) that C satisfies the following properties.

15.14. Let C_0 denote the maximal subcomplex of C such that $F^{-q}(|C_0|) \subset |C|$.

(a) $F^q(|C|) \subset \operatorname{Int}(|C|)$, $F^q(|C_0|) \subset \operatorname{Int}(|C_0|)$.

(b) The maximal diameter of any cell in the cell structures $C, F^q(C)$, $F^{-q}(C)$ is very much less than the distance from $F^q(|C|)$ (or from $F^{2q}(|C|)$) to the topological boundary for $|C|$ (or to the topological boundary for $F^q(|C|)$).

We define a sequence $C_1, C_2, C_3, \ldots$ of cellular subdivisions of C_0 as follows. We proceed by induction, assuming that the $C_0, C_1, C_2, \ldots$, C_{i-1} have already been defined. Now define C_i to be the minimal cellular subdivision of C_{i-1} satisfying the following properties.

(a) If $e \in C_{i-1}$ and $e \not\subset F^{iq}(|C|)$ then e is also a cell in C_i.

(b) If $e \in C_{i-1}$ and $e \subset F^{iq}(|C|)$ then the restriction of C_i to e is gotten by intersecting e with all the cells in the cell structures $F^{iq}(C)$ and $F^{-iq}(C)$.

Note that it follows from 1.3, 15.14 that for each $i \geq 1$ C_i is a well defined subdivision of C_{i-1} satisfying the following properties.

15.15. (a) For each $i \geq 1$ and each cell $e \in C_i$ we have that $F^q(e)$ is the underlying subset of a subcomplex of C_{i+1}.

(b) Let E denote the collection of all cells e such that for some $i \geq 1$ (depending on e) we have that $e \in C_j$ for all $j \geq i$. Then E is a cell structure for the set $|C_0| - \Lambda$.

(c) Given $\varepsilon > 0$ there is $i \geq 1$ such that for any cell $e \in C_i$ which is not a cell of C_{i-1} the diameter of e is less than ε.

(d) For each cell $e \in E$, $F^q(e)$ must be the underlying set of a subcomplex of E. (This is implied by (a) and (b).)

We call the sequence of subdivisions $C_1, C_2, C_3, \ldots$ the **dynamical subdivision process** for the Markov cell structure C. The **dynamical subdivision process pairings** are mappings $\Phi_u : D \times D \to Z_2$, $\Phi_s : D \times D \to Z_2$, $\Psi : D \times D \to Z_2$, $\Phi : D \times D \to Z_2$ defined as follows. Set $D = \cup_{i \geq 1} C_i$. For any $e_1, e_2 \in D$ set $\Phi_u(e_1, e_2) = 1$ if $e_1 \subset \partial_u e_2$; otherwise set $\Phi_u(e_1, e_2) = 0$. Set $\Phi_s(e_1, e_2) = 1$ if $e_1 \subset \partial_s e_2$; otherwise set $\Phi_s(e_1, e_2) = 0$. Set $\Psi(e_1, e_2) = 1$ if $(e_1 - \partial e_1) \cap F^q(e_2 - \partial e_2) \neq \emptyset$; otherwise set $\Psi(e_1, e_2) = 0$. Set $\Phi(e_1, e_2) = 1$ if $e_1 \subset e_2$; set $\Phi(e_1, e_2) = 0$ otherwise.

We leave the verification of the following claim to the reader.

CLAIM 15.16. The dynamical subdivision pairings $\Phi : D \times D \to Z_2$, $\Phi_u : D \times D \to Z_2$, $\Phi_s : D \times D \to Z_2$, $\Psi : D \times D \to Z_2$ can be computed from the Markov pairings $\phi_u : C \times C \to Z_2$, $\phi_s : C \times C \to Z_2$, $\psi : C \times C \to Z_2$.

We shall complete the proof of theorem 15.8 in the following three steps.

STEP I. In this step we will construct from the dynamical subdivision process pairings Φ_u, Φ_s, Φ, Ψ a sequence $C'_1, C'_2, C'_3, \ldots$ of finite regular cell complexes, and for each $i \geq 1$ and each $e \in C'_i$ we will construct two subcomplexes $\partial_u e$ and $\partial_s e$ of C'_i, all satisfying the following properties.

15.17. (a) C'_{i+1} is a subdivision of C'_i for each $i \geq 1$.

(b) For each $i \geq 1$ there is a homeomorphism $h_i : |C'_i| \to |C_i|$ which maps each cell of C'_i onto a cell of C_i.

(c) For any $j \geq i \geq 1$ and any cell $e \in C'_i$ we have that $h_i(e) = h_j(e)$, $\partial_u h_i(e) = h_i(\partial_u e)$, and $\partial_s h_i(e) = h_i(\partial_s e)$.

(d) The $h_i : |C'_i| \to |C_i|$ are uniquely determined by the $\Phi, \Phi_u, \Phi_s, \Psi$ up to isotopy, where each level of the isotopy must satisfy parts (a)(b)(c).

For each $i \geq 1$ we can construct C'_i from the pairings $\Phi_u : C_i \times C_i \to Z_2$, $\Phi_s : C_i \times C_i \to Z_2$ as follows. Note that since C_i is a regular cell complex we can determine the dimension of any cell $e \in C_i$ from these pairings. The construction of C'_i proceeds by induction over the dimension of the cells of C_i. Suppose that the $(r-1)$-skeleton $(C'_i)^{r-1}$ has already been constructed so as to satisfy the following properties.

15.18$(r-1)$. For each cell $e \in C_i^{r-1}$ there is a corresponding cell $e' \in (C'_i)^{r-1}$. If $\partial_u e = \cup_{i \in I_u} e_i$ then $\partial_u e' = \cup_{i \in I_u} e'_i$; if $\partial_s e = \cup_{i \in I_s} e_i$ then $\partial_s e' = \cup_{i \in I_s} e'_i$.

To get $(C'_i)^r$ we must attach an r-cell e' to $(C'_i)^{r-1}$ for each r-cell $e \in C^r$. Write $\partial e = \cup_{i \in I} e_i$ and set $\partial e' = \cup_{i \in I} e'_i$, $e' = \text{cone}(\partial e')$.

Set $\partial_u e' = \cup_{i\in I_u} e'_i$ and $\partial_s e' = \cup_{i\in I_s} e'_i$, where $\partial_u e = \cup_{i\in I_u} e_i$ and $\partial_s e = \cup_{i\in I_s} e_i$. Note that $(C'_i)^r$ will be a well defined regular cell complex satisfying 15.18(r) provided that each $\partial e'$ is an $(r-1)$-sphere. This can be deduced from 15.18($r-1$) and 15.19 because each of the ∂e is an $(r-1)$-sphere.

15.19. Let K, L denote two finite regular cell complexes. Suppose that there is a one-one correspondence (denoted $e \leftrightarrow e'$) between the cells of K and L such that for each $e \in K$ if $\partial e = \cup_{i\in I} e_i$ (with the e_i in K) then we must have $\partial e' = \cup_{i\in I} e'_i$. Then there is a homeomorphism $h : |K| \to |L|$ such that $h(e) = e'$ for each $e \in K$.

Note that the construction of $h : |K| \to |L|$ in 15.19 proceeds by induction over the dimension of the cells in K. Once $h : |K^{r-1}| \to |L^{r-1}|$ has been constructed we can extend it to $h : |K^r| \to |L^r|$ by applying Alexander's trick to each r-cell of K^r.

By applying 15.19 to each pair C'_i, C_i we get homeomorphisms $g_i : |C'_i| \to |C_i|$ which map each cell of C'_i onto a cell of C_i. For any $j \geq i \geq 1$ the images of the cells of C_i under the map g_j^{-1} give to $|C'_j|$ a cell structure which we denote by $C'_{i,j}$. Note that C'_j is a subdivision of $C'_{i,j}$ and that $C'_{i,j}$ can be constructed from C'_j and the pairing $\Phi : C_j \times C_i \to Z_2$. In fact for $e \in C_i$ we have that $g_j^{-1}(e) = \cup_t e'_t$ where the union runs over all $e'_t \in C'_j$ such that $\Phi(e_t, e) = 1$. Now choose (by 15.18 and 15.19) any homeomorphism $g_{i,j} : |C'_{i,j}| \to |C'_i|$ which maps the cell $\cup_t e'_t$ onto the cell e' (for every pair of such cells). Note that $g_{i,j}$ embeds C'_j as a cellular subdivision of C'_i. So we may first apply this argument with $i = 1$, $j = 2$ to get C'_2 to be a subdivision of C'_1; next apply the argument to get C'_3 to be a subdivision for C'_2; etc. In this way we can arrange that the $C'_1, C'_2, C'_3, \ldots$ satisfy 15.17(a). Now we can apply 15.19 to each pair C'_i, C_i to get the homeomorphisms $h_i : |C'_i| \to |C_i|$ of 15.17(b)(c).

This completes step I.

STEP II. In this step we construct a topological space X and a mapping $g : X \to X$ using only the information contained in the dynamical subdivision process pairings Φ, Φ_u, Φ_s, Ψ. In step III below we will show that $g : X \to X$ is topologically conjugate to the map $F^q : |C_0| \to |C_0|$.

Consider the collection E' of cells e in $\cup_{i=1}^{\infty} C'_i$ satisfying: there is $i \geq 1$ (depending on e) such that for every $j \geq i$ we have $e \in C'_j$. It follows from 15.15(b) and 15.17 that E' is a cell structure for a region $|E'|$ in $|C'_1|$. Note it follows from 15.19 that there is a map $r : |E'| \to |E|$ satisfying the following properties.

15.20. (a) r is a cellular homeomorphism from E' to E.

(b) The isotopy class of r is uniquely determined by the dynamical subdivision process pairings. (Here we require that each level of the isotopy satisfy (a).)

Let $s : |E'| \to |E'|$ denote any map satisfying the following properties.

15.21. (a) $s : |E'| \to |E'|$ is an embedding.
(b) For each $e \in E'$ we have that $r \circ s(e) = F^q \circ r(e)$. In particular it follows that for each cell $e \in E'$ the image $s(e)$ is the underlying set of a subcomplex of E'.

Note that $s : |E'| \to |E'|$ of 15.21 is uniquely determined (up to isotopy class) by the dynamical subdivision process pairings. (Here we require that each level of the isotopy satisfy 15.21(a)(b).)

We will now define the space X to be a quotient space $|C_1'|/\sim$ of $|C_1'|$. Set $x \sim y$ if and only if either $x = y$ or if for any given $i \geq 1$ there is $j \geq i$ and two cells $e_1, e_2 \in C_j$ satisfying the following properties: $e_1 \cap e_2 \neq \emptyset$; $x \in e_1$, $y \in e_2$; $e_1 \notin C_i'$, $e_2 \notin C_i'$. Note that the quotient map $q : |C_1'| \to X$ satisfies the following property.

15.22. $q : |E'| \to q(|E'|)$ is a homeomorphism.

Finally we can define the map $g : X \to X$. Set $g|q(|E'|)$ equal to the composition of maps $q \circ s \circ (q^{-1}|q(|E'|))$, where s comes from 15.21. To define $g(x)$ for any $x \in q(|C_1'| - |E'|)$ choose $x' \in C_1'$ such that $q(x') = x$ and choose for each $i \geq 1$ a cell $e_i \in C_i'$ such that $x' \in e_i$. Next choose $y' \in \bigcap_{i=1}^{\infty} h_i^{-1} \circ F^q \circ h_i(e_i)$, and set $g(x) = q(y')$.

It is left to the reader to check that $g : X \to X$ is a well defined map whose construction depends only on the dynamical subdivision process pairings.

This completes step II.

STEP III. In this step we construct a homeomorphism $h : X \to |C_0|$ such that $h \circ g = F^q \circ h$ holds on X.

For any $x \in q(|C_1'| - |E'|)$ we define $h(x)$ as follows. Choose $x' \in C_i'$ such that $q(x') = x$. Choose cells $e_i' \in C_i'$, $i = 1, 2, \ldots$, such that $x' \in e_i'$. Let $e_i \in C_i$ denote the corresponding cells of the C_i, $i = 1, 2, \ldots$ (corresponding under the h_i of 15.17). Note that it follows from 15.15(b)(c) that the intersection $\bigcap_{i=1}^{\infty} e_i$ is a single point denoted by y. Set $h(x) = y$.

We define $h|q(|E'|)$ to be the composition of maps $q \circ T \circ (q^{-1}|q(|E'|))$ where $T : |E'| \to |E|$ is defined by induction over the skeleta of E' as follows. Suppose that $T : |E'^i| \to |E^i|$ has been defined so as to satisfy the following properties. (Note, by 15.22, $h|q(|E'|)$ will be well defined if T is well defined.)

15.23(i). (a) $T : |E'^i| \to |E^i|$ is a cellular homeomorphism from E'^i to E^i.

(b) $T \circ s(x) = F^q \circ T(x)$ holds for all $x \in |E'^i|$, where $s : |E'| \to |E'|$ comes from 15.21.

(c) For each $e \in E'^i$ choose $j \geq 1$ such that $e \in C'_j$. Then $h_j(e) = T(e)$, where $h_j : |C'_j| \to |C_j|$ comes from 15.17.

Note that there is a collection S of $(i+1)$-dimensional cells in E' satisfying the following property.

15.24. (a) $\cup s^j(e) = |E'^{i+1}|$, where the union runs over all $e \in S$ and $j = 0, 1, 2, \ldots$.

(b) For any two distinct $e_1, e_2 \in S$ and any $j \geq 0$ we always have that $(e_1 - \partial e_1) \cap s^j(e_2) = \emptyset$.

For each $e' \in S$ let $e \in E$ denote the corresponding cell in E (corresponding under the map $r : |E'| \to |E|$ of 15.20). Extend $T : \partial e' \to \partial e$ in any way to a homeomorphism $T : e' \to e$. Now for any other $(i+1)$-dimensional cell e' in $E' - S$ there is by 15.24 an integer $j > 0$ and a cell $e'_1 \in S$ such that $s^{-j}(e') \subset e'_1$. Define $T : e' \to E$ to be the composition of maps $F^{jq} \circ (T|e'_1) \circ (s^{-j}|e')$. Note that $T : |E'^{i+1}| \to E^{i+1}|$ satisfies 15.23($i+1$). This completes the inductive description of T.

It is left to the reader to verify that $h : X \to |C_0|$ just defined is a well defined homeomorphism satisfying $h \circ g = F^q \circ h$ for $g : X \to X$ defined as in step II.

This completes step III and the proof of theorem 15.8.

REFERENCES

[1] R. L. Adler and B. Weiss, *Entropy: a complete metric invariant for automorphisms of the torus,* Proc. Nat. Acad. Sci. USA 57 (1967), 1573–1576.

[2] D. V. Anosov, *Geodesic flows on closed Riemannian manifolds of negative curvature,* Trudy Mat. Inst. Steklov 90 (1967).

[3] E. Bothe, *An example of a strange basic set,* preprint.

[4] R. Bowen, *On Axiom A diffeomorphisms,* C.M.B.S. Regional Conference Series in Math., no. 38, Amer. Math. Soc. (1978).

[5] R. Bowen, *Markov partitions for Axiom A diffeomorphisms,* Amer. J. of Math. 92 (1970), 725–747.

[6] F. T. Farrell and L. E. Jones, *Markov cell structures,* Bull. Amer. Math. Soc. 83 (1977), 739–740.

[7] F. T. Farrell and L. E. Jones, *Markov cell structures for expanding maps in dimension two,* Trans. Amer. Math. Soc. 255 (1979), 315–327.

[8] F. T. Farrell and L. E. Jones, *Expanding immersions on branched manifolds,* Amer. J. of Math. 103 (1981), 41–101.

[9] J. Franks, *Homology and dynamical systems,* C.B.M.S. Regional Conference in Math., no. 49, Amer. Math. Soc. (1982).

[10] J. Guckenheimer, *Endomorphisms of the Riemannian sphere,* Proc. Symp. in Pure Math., vol. 14, Amer. Math. Soc. (1970), 95–123.

[11] M. Hirsch and C. Pugh, *Stable manifolds and hyperbolic sets,* Proc. Symp. in Pure Math., vol. 14, Amer. Math. Soc. (1970), 133–164.

[12] J. F. P. Hudson, *Piecewise linear topology,* W. A. Benjamin Inc., New York and Amsterdam (1969).

[13] L. E. Jones, *Locally strange hyperbolic sets,* Trans. Amer. Math. Soc. 275 (1983), 153–162.

[14] L. E. Jones, *Anosov diffeomorphisms and expanding immersions: I and II,* Trans. Amer. Math. Soc. 289 (1985), 115–131, and Trans. Amer. Math. Soc. 294 (1986), 197–216.

[15] K. Krzyzewski, *On a connection between expanding mappings and Markov chains,* Bull. Acad. Polon. Sci. Ser. Sci. Math. Astron. Phys. 19 (1971), 291–293.

[16] J. Milnor, *On the existence of a connection with curvature zero,* Comment. Math. Helv. 32 (1958), 215–223.

[17] J. Milnor, *Whitehead torsion,* Bull. Amer. Math. Soc. 72 (1966), 358–426.

[18] J. R. Munkres, *Elementary differential topology,* Princeton University Press, Princeton, N.J. (1963).

[19] F. Przyztycki, *Construction of invariant sets for Anosov diffeomorphisms and hyperbolic attractors,* preprint.

[20] J. Robbin, *A structure stability theorem,* Ann. of Math. 94 (1971), 477–493.

[21] C. Robinson, *Structural stability of C^1-diffeomorphisms,* J. Diff. Equations 22 (1976), 28–73.

[22] C. Robinson and R. F. Williams, *Classification of expanding attractors: an example,* Topology 15 (1976), 321–323.

[23] M. Shub, *Endomorphisms of compact differentiable manifolds,* Amer. J. of Math. 91 (1969), 175–199.

[24] M. Shub and D. Sullivan, *Homology and dynamical systems,* Topology 14 (1975), 109–132.

[25] I. G. Sinai, *Markov partitions and C-diffeomorphisms,* Funkcional. Anal. i Prilozen 2 (1968), 64–89.

[26] S. Smale, *Differential dynamical systems,* Bull. Amer. Math. Soc. 73 (1967), 747–818.

[27] D. Sullivan and R. F. Williams, *On the homology of attractors,* Topology 15 (1976), 259–262.

[28] R. F. Williams, *Expanding Attractors,* Inst. Hautes Etudes Sci. Publ. Math. 43 (1974), 169–203.

Editorial Information

To be published in the *Memoirs*, a paper must be correct, new, nontrivial, and significant. Further, it must be well written and of interest to a substantial number of mathematicians. Piecemeal results, such as an inconclusive step toward an unproved major theorem or a minor variation on a known result, are in general not acceptable for publication. *Transactions* Editors shall solicit and encourage publication of worthy papers. Papers appearing in *Memoirs* are generally longer than those appearing in *Transactions* with which it shares an editorial committee.

As of March 1, 1993, the backlog for this journal was approximately 7 volumes. This estimate is the result of dividing the number of manuscripts for this journal in the Providence office that have not yet gone to the printer on the above date by the average number of monographs per volume over the previous twelve months, reduced by the number of issues published in four months (the time necessary for preparing an issue for the printer). (There are 6 volumes per year, each containing at least 4 numbers.)

A Copyright Transfer Agreement is required before a paper will be published in this journal. By submitting a paper to this journal, authors certify that the manuscript has not been submitted to nor is it under consideration for publication by another journal, conference proceedings, or similar publication.

Information for Authors

Memoirs are printed by photo-offset from camera copy fully prepared by the author. This means that the finished book will look exactly like the copy submitted.

The paper must contain a *descriptive title* and an *abstract* that summarizes the article in language suitable for workers in the general field (algebra, analysis, etc.). The *descriptive title* should be short, but informative; useless or vague phrases such as "some remarks about" or "concerning" should be avoided. The *abstract* should be at least one complete sentence, and at most 300 words. Included with the footnotes to the paper, there should be the 1991 *Mathematics Subject Classification* representing the primary and secondary subjects of the article. This may be followed by a list of *key words and phrases* describing the subject matter of the article and taken from it. A list of the numbers may be found in the annual index of *Mathematical Reviews*, published with the December issue starting in 1990, as well as from the electronic service e-MATH [**telnet e-MATH.ams.org** (or **telnet 130.44.1.100**). Login and password are **e-math**]. For journal abbreviations used in bibliographies, see the list of serials in the latest *Mathematical Reviews* annual index. When the manuscript is submitted, authors should supply the editor with electronic addresses if available. These will be printed after the postal address at the end of each article.

Electronically-prepared manuscripts. The AMS encourages submission of electronically-prepared manuscripts in $\mathcal{A}\mathcal{M}\mathcal{S}$-TeX or $\mathcal{A}\mathcal{M}\mathcal{S}$-LaTeX. To this end, the Society has prepared "preprint" style files, specifically the amsppt style of $\mathcal{A}\mathcal{M}\mathcal{S}$-TeX and the amsart style of $\mathcal{A}\mathcal{M}\mathcal{S}$-LaTeX, which will simplify the work of authors and of the production staff. Those authors who make use of these style files from the beginning of the writing process will further reduce their own effort.

Guidelines for Preparing Electronic Manuscripts provide additional assistance and are available for use with either $\mathcal{A}_{\mathcal{M}}\mathcal{S}$-TEX or $\mathcal{A}_{\mathcal{M}}\mathcal{S}$-LATEX. Authors with FTP access may obtain these *Guidelines* from the Society's Internet node e-MATH.ams.org (130.44.1.100). For those without FTP access they can be obtained free of charge from the e-mail address guide-elec@math.ams.org (Internet) or from the Publications Department, P. O. Box 6248, Providence, RI 02940-6248. When requesting *Guidelines* please specify which version you want.

Electronic manuscripts should be sent to the Providence office only after the paper has been accepted for publication. Please send electronically prepared manuscript files via e-mail to pub-submit@math.ams.org (Internet) or on diskettes to the Publications Department address listed above. When submitting electronic manuscripts please be sure to include a message indicating in which publication the paper has been accepted.

For papers not prepared electronically, model paper may be obtained free of charge from the Editorial Department at the address below.

Two copies of the paper should be sent directly to the appropriate Editor and the author should keep one copy. At that time authors should indicate if the paper has been prepared using $\mathcal{A}_{\mathcal{M}}\mathcal{S}$-TEX or $\mathcal{A}_{\mathcal{M}}\mathcal{S}$-LATEX. The *Guide for Authors of Memoirs* gives detailed information on preparing papers for *Memoirs* and may be obtained free of charge from AMS, Editorial Department, P. O. Box 6248, Providence, RI 02940-6248. The *Manual for Authors of Mathematical Papers* should be consulted for symbols and style conventions. The *Manual* may be obtained free of charge from the e-mail address cust-serv@math.ams.org or from the Customer Services Department, at the address above.

Any inquiries concerning a paper that has been accepted for publication should be sent directly to the Editorial Department, American Mathematical Society, P. O. Box 6248, Providence, RI 02940-6248.

Editors

Recent Titles in This Series

(*Continued from the front of this publication*)

(See the AMS catalog for earlier titles)